BIOCHEMISTRY AND MOLECULAR BIOLOGY IN THE POST GENOMIC ERA

INTERNAL ROTATION IN MOLECULAR CRYSTALS

COMPUTATIONAL STUDIES

Biochemistry and Molecular Biology in the Post Genomic Era

Additional books in this series can be found on Nova's website under the Series tab.

Additional e-books in this series can be found on Nova's website under the e-book tab.

BIOCHEMISTRY AND MOLECULAR BIOLOGY IN THE POST GENOMIC ERA

INTERNAL ROTATION IN MOLECULAR CRYSTALS

COMPUTATIONAL STUDIES

XIANLONG WANG
Key Laboratory for NeuroInformation of Ministry of Education, School of Life Science and Technology, University of Electronic Science and Technology of China

nova publishers
New York

Library of Congress Cataloging-in-Publication Data

ISBN: 978-1-63482-214-5

Library of Congress Control Number: 2015932299

Published by Nova Science Publishers, Inc. † New York

CONTENTS

PREFACE

Molecular rotors, motors and machines are a major subject of research in the current era of nanotechnology. Complex biological motors, which fuel most cellular processes in living organisms, are found throughout the natural world. Chemists design and synthesize molecules with the hope to perform useful work as biological systems do. For example, the research group of Professor James Tour at Rice University created the first molecular nanocar in 2005. A lot of pioneering studies have been also carried out in the research groups of Miguel A. Garcia-Garibay at the University of California, Los Angles, Ben L. Feringa at the University of Groningen, Josef Michl at the University of Colorado at Boulder and others. Internal rotation is a key component in the design of artificial molecular rotors and it is ubiquitous. Although theoretical studies on the internal rotation dated back to 1930's (Kenneth S. Pitzer discussed the internal rotation around the single bond in ethane in 1936), there remain many challenging problems in designing molecular rotors, such as rotational frequency, assembling and unidirectionality. The author has been focused on the study the internal rotation of small rotors, such as methyl, trifluoromethyl, *tert*-butyl and methyl groups, in molecular crystals in the past decade. In studying molecular crystals, the large system size and the accurate description of intermolecular interaction, which is dominated by the London's dispersion, are two major difficulties. The author has developed a cluster model approach to mimic the micro-environment of molecular rotors in the crystalline solids and it has been successfully to study the rotational dynamics in many aromatic molecular systems. In this monograph, the author presents the details of the studies on a few representative *tert*-butyl aromatic derivatives and phenol compounds. Through the comparative studies in both isolated molecules and clusters, the role of *intra*molecular and *inter*molecular interactions was analyzed. The rotational coupling problem was investigated through the two-dimensional potential energy surfaces. The computational barrier heights were found to be in good agreement with the activation energy measured in the variable temperature nuclear magnetic resonance relaxometry experiments.

Xianlong Wang, PhD
Associate Professor
School of Life Science and Technology
University of Electronic Science and Technology of China

ACKNOWLEDGMENT

The author acknowledges the financial support from the National Natural Science Foundation of China (Grant No. 21103016) and the Key Laboratory for NeuroInformation of Ministry of Education at the University of Electronic Science and Technology of China.

Chapter 1

INTRODUCTION

1.1. MOTIVATION

Internal rotation dynamics around a single bond has been a long-pursued topic of chemical physics since the internal rotation of methyl groups about the C-C bond in ethane was found to have a significant barrier in the1930's [1-4]. Many experimental investigations have been carried out on internal rotation of various kinds of rotors in a wide range of molecular systems [5-14]. At the same time, many theoretical works are also devoted to looking at various problems related to internal rotation [15-32]. The study of internal rotation helps us to understand some fundamental questions of molecular systems, *e.g.* molecular structure, *intra*molecular and *inter*molecular interactions, energy transfer and structural relaxation.

In a *t*-butyl group, there are four degrees of freedom associated with internal rotation. These are overall rotation of the *t*-butyl group and internal rotation of the three constituent methyl groups. Experimental studies on the internal rotation of *t*-butyl groups have been carried out in crystalline solids and in inclusion compounds using nuclear magnetic resonance relaxometry method [33-43]. However, there are many questions that remain unanswered. Several key questions include the following. (1) How many distinct internal rotation processes are there for a *t*-butyl group in these solid compounds? (2) What is the relationship between the overall rotation of the *t*-butyl group and the internal rotation of the constituent methyl groups? Are they coupled with each other or independent? (3) How does molecular structure affect the internal rotation? (4) Experimental results have shown that the crystal packing environments have a great influence on *t*-butyl group rotation. But how? What are the details concerning the role of the intermolecular interactions in the *t*-butyl rotational processes? (5) How do the *inter*molecular interactions and the *intra*molecular interactions interplay in the internal rotation processes in the solid state? (6) In macroscopic solid samples, there are many *t*-butyl groups and their rotation axes are relatively fixed. Are their rotations coupled or independent? To answer these questions, computational studies are complementary to the experimental methods. Although computational studies on molecular systems in the gas phase can be found in the literature [44-46], computational studies in the solid state have yet to be done. This is the motive for the study reported in this work. The work represents the first of its kind on *t*-butyl internal rotation problems in the crystalline solids.

Experimental methods used to study internal rotation in the gas phase include microwave spectroscopy [6, 47-48], vibration spectroscopy [48-49] and other molecular spectroscopic techniques. Thermal properties have also been used for this purpose [1]. Supersonic laser jet spectroscopy can be used to determine the favorable orientation of the rotor groups in the isolated molecules and in molecular clusters [8, 50-57]. In solution, NMR spectroscopy is often used to measure rotational barriers [9-14, 58-63]. The *J*-coupling among spins, which are possibly modulated by the internal rotation, is the main relaxation mechanism. But in the solid state, the spin dipolar interaction is the main relaxation mechanism. If the spin dipolar interaction is modulated by internal rotation, the rotational dynamics can be investigated from the spin relaxation rate results. See the literature [64-65] for detailed background theory. The effective reorientation barriers for *t*-butyl and methyl groups, measured using the NMR relaxometry technique, will be compared with the computational results in this work. X-ray diffraction and neutron diffraction experiments also show the capability in measuring rotational barriers in crystalline solids [66-67].

To theoretically study the internal rotation in crystalline solids, we use cluster models, built from the X-ray crystal structures, to mimic the crystal packing environment for the rotor groups. Cluster models have been used extensively to study chemical reactions in the solid state and the defect properties in the crystal solids [68]. They also proved successful in simulating the local environment of the *t*-butyl groups in this work.

The molecular systems studied in this work include three aromatic hydrocarbons (3-*t*-butylchrysene, 2,6-di-*t*-butylnaphthalene and 1,4-di-*t*-butylbenzene), two phenols (2-*t*-butyl-4-methylphenol and 2,6-di-*t*-butyl-4-methylphenol), and 3-(trifluoromethyl)phenanthrene. The three aromatic hydrocarbons lack strong intramolecular steric effects on *t*-butyl group rotation but the sizes of the aromatic rings change systematically. In the two phenols, there are hydroxyl groups on the *ortho*-positions of the *t*-butyl groups, which may exert both steric and electronic effects on the internal rotations of the *t*-butyl groups. The study on the trifluoromethyl group in 3-(trifluoromethyl)phenanthrene provides an opportunity for comparisons with the internal rotation of methyl and *t*-butyl groups on aromatic systems. Experimentally, the system is also interesting because both the proton-1 and fluorine-19 relaxation results can be used to detect the rotational dynamics [69-70].

The structure of this book is organized in the following manner. The internal rotations of the *t*-butyl groups and the constituent methyl groups in the three aromatic compounds are studied for the isolated molecules in Chapter 2 and in the crystalline solids in Chapter 3. Results on the internal rotations of the *t*-butyl groups, the constituent methyl groups and the hydroxyl groups on the two phenols are reported for the isolated molecules in Chapter 4 and in the crystalline solids in Chapter 5. Results on the internal rotation of the trifluoromethyl group in 3-(trifluoromethyl)phenanthrene are reported for both the isolated molecule and for the crystalline solid in Chapter 6. The Schrödinger equations of one-dimensional rotors and two-dimensional rotors are solved in the Appendix and the results are discussed in connection with dynamical models for methyl and *t*-butyl groups.

1.2. REFERENCES

[1] Pitzer, KS. *Molecular structure and statistical thermodynamics: selected papers of Kenneth S. Pitzer*. World Scientific: Singapore, 1993.

[2] Mizushima, S. *Structure of molecules and internal rotation*. Academic Press: New York, 1954.

[3] Goodman, L; Pophristic, V; Weinhold, F. Origin of methyl internal rotation barriers. *Acc. Chem. Res.*, 1999, 32(12), 983-993.

[4] Pophristic, V; Goodman, L. Hyperconjugation not steric repulsion leads to the staggered structure of ethane. *Nature*, 2001, 411, 565-568.

[5] Horsewill, A. J. Quantum tunneling aspects of methyl group rotation studied by NMR. *Prog. Nucl. Magn. Reson. Spectroscopy*, 1999, 35, 359-389.

[6] Lin, CC; Swalen, JD. Internal rotation and microwave spectroscopy. *Rev. Modern Phys.*, 1959, 31(4), 841.

[7] Pottage, JT; Flower, DR; Davis, SL. The torsional excitation of methanol by helium. *J. Phys. B*, 2004, 37, 165-177.

[8] Spangler, LH; Pratt, DW. Internal rotation dynamics from electronic spectroscopy in supersonic jets and beams. In *Jet spectroscopy and molecular dynamics*, Hollas, J. M. and Phillips, D., Eds. Blackie Academic & Professional: New York, 1995, 366-399.

[9] Dalling, DK; Ladner, KH; Grant, DM; Woolfenden, WR. Carbon-13 Magnetic Resonance. 27. The dependence of chemical shifts on methyl rotational conformations and dynamics in the methylated benzenes and naphthalenes. *J. Am. Chem. Soc.*, 1977, 99(22), 7142-7150.

[10] Jarek, RL; Flesher, RJ; Koo, SS. Kinetics of internal rotation of *N,N*-dimethylacetamide: A spin-saturation transfer experiment. *J. Chem. Ed.*, 1997, 74 (978), 982.

[11] Lipari, G; Szabo, A. Nuclear magnetic resonance relaxation in nucleic acid fragments: Models for internal motion. *Biochemistry*, 1981, 20, 6250-6256.

[12] Parr, WJE; Schaefer, T. A proton magnetic resonance determination of the barrier to internal rotation in benzethiol in solution. *J. Magn. Reson.*, 1977, 25, 171-176.

[13] Schaefer, T; Lee, LB. An H-1 NMR conformational analysis of 3-phenylpentane in solution. *Can. J. Chem.*, 1992, 71, 520-524.

[14] Schaefer, T; Parr, WJE; Danchura, W. Concerning the barrier to internal rotation in isopropylbenzene in solution. *J. Magn. Reson.*, 1977, 25, 167-170.

[15] Nielsen, HH. The torsion oscillator-rotator in the quantum mechanics. *Phys. Rev.*, 1932, 40, 445-456.

[16] Koehler, JS; Dennison, DM. Hindered rotation in methyl alcohol. *Phys. Rev.*, 1940, 57, 1006-1021.

[17] Mellor, WE; Lee, AR; Kalotas, TM. Alternative calculations for internal rotations: Assessment via Mathieu and multi-Fourier term potentials. *Chem. Phys.*, 1997, 219, 257-263.

[18] Spangler, LH. Structural information from methyl internal rotation spectroscopy. *Ann. Rev. Phys. Chem.*, 1997, 48, 481-510.

[19] Chen, CC; Bozzelli, JW. Structures, Intramolecular rotation barriers, and thermochemical properties of methyl, ethyl, methyl isopropyl, and methyl *tert*-butyl ethers and the corresponding radicals. *J. Phys. Chem. A*, 2003, 107, 4531-4546.

[20] Gamble, WL; Miyagawa, I; Hartman, RL. Some effects of quantization of internal rotation on spin-lattice relaxation and hyperfine structure. *Phys. Rev. Lett.*, 1967, 20(9), 415-418.

[21] Allen, L. C. Energy component analysis of rotational barriers. *Chem. Phys. Lett.*, 1968, 2(8).

[22] Benassi, R; Taddei, F. A Theoretical approach to the factorization of the effects governing the barrier for internal rotation around the C(sp2)-C(sp3) bond into α-substituted toluenes. *J. Mol. Struct. (Theochem)*, 1997, 418, 59-71.

[23] Brunelle, JA; Ketendre, LJ; Weltin, EE; Brown, JH; Bushweller, C. H. Barriers to internal rotation in CF_3CFCl_2, CF_3CHCl_2, CF_3CF_2Br, CF_3CF_2Cl, CF_3CH_2Br and CF_3CH_2Cl. F-19 and H-1 dynamic NMR studies. *Ab initio* molecular orbital barrier calculations. *J. Phys. Chem.*, 1992, 96, 9225-9232.

[24] Cacelli, I; Prampolini, G. Torsional barriers and correlations between dihedrals in *p*-polyphenyls. *J. Phys. Chem. A*, 2003, 107, 8665-8670.

[25] Halpern, AM; Glendening, ED. An intrinsic reaction coordinate calculation of the torsional potential in ethane: comparison of the computationally and experimentally derived torsional transitions and the rotational barrier. *J. Chem. Phys.*, 2003, 119 (21), 11186-11191.

[26] Mason, EA; Kreevoy, MM. A simple model for barriers to internal rotation. *J. Am. Chem. Soc.*, 1955, 77, 5808-5814.

[27] Palmer, M., H; Camp, PJ. An *ab initio* study of vibration and rotation effects in isocyanatometane, isothiocynantomethane and azidomethane. *Mol. Phys.*, 2003, 101(20), 3053-3062.

[28] Radom, L; Lathan, WA; Hehre, WJ; Pople, JA. Molecular orbital theory of the electronic structure of organic compounds. XVII. Internal rotation in 1,2-disubstituted ethanes. *J. Am. Chem. Soc.*, 1973, 95(3), 693-698.

[29] Schaefer, T; Sebastian, R; Penner, GH. Theoretical and experimental data on the internal rotational potential in isopropylbenzene. *Can. J. Chem.*, 1987, 66, 1495-1499.

[30] Sztraka, L; Csonka, GI. Flexible *ab initio* geometry of methylamine and its internal rotation. *J. Mol. Struct*, 1997, 410-411, 387-390.

[31] Yanborisov, VM; Zaitesev, MY; Monakov, YB. Determination of internal rotation barrier. *Russian J. App. Chem.*, 2001, 74(12), 1942-1944.

[32] Zierkiewicz, W; Michalska, D; Hobza, P. The barrier to internal rotation and electronic effects in *para*-halogenophenols: theoretical study. *Chem. Phys. Lett.*, 2004, 386, 95-100.

[33] Beckmann, PA; Burbank, KS; Clemo, KM; Slonaker, EN; Averill, K; Dybowski, C; Figueroa, JS; Glatfelter, A; Koch, S; Liable-Sands, LM; Rheingold, AL., 1H nuclear magnetic resonance spin-lattice relaxation, ^{13}C magic-angle-spinning nuclear magnetic resonance spectroscopy, differential scanning calorimetry, and X-ray diffraction of two polymorphs of 2,6-di-*tert*-butyl-naphthalene. *J. Chem. Phys.*, 2000, 113, 1958-1965.

[34] Beckmann, PA; Buser, CA; Gullifer, K; Mallory, FB; Mallory, CW; Rossi, GM; Rheingold, AL. Methyl and *t*-butyl group reorientation in planar aromatic solids: Low-

frequency nuclear magnetic resonance relaxometry and x-ray diffraction. *J. Chem. Phys.*, 2003, 118(24), 11129-11138.

[35] Beckmann, PA; Fusco, FA; O'Neill, AE. Proton spin-lattice relaxation and intramolecular reorientation in solids. *J. Magn. Reson.*, 1984, 59, 63-70.

[36] Beckmann, PA; Paty, C; Allocco, E; Herd, M; Kuranz, C; Rheingold, AL. The relationship between crystal structure and methyl and *t*-butyl group dynamics in van der Waals organic solids. *J. Chem. Phys.*, 2004, 120(11), 5309-5314.

[37] Beckmann, PA; Ratcliffe, CI; Dunell, B. Proton spin-lattice relaxation and methyl group rotation. *J. Magn. Reson.*, 1978, 32, 391-402.

[38] Conn, KG; Beckmann, PA; Mallory, CW; Mallory, FB. Methyl reorientation in methylphenanthrenes. I. Solid state proton spin-lattice relaxation in the 3-methyl, 9-methyl and 3,9-dimethyl systems. *J. Chem. Phys.*, 1987, 87 (1), 20-27.

[39] Rheingold, AL; Figueroa, JS; Dybowski, C; Beckmann, PA. Superlattices, polymorphs and solid-state NMR spin–lattice relaxation (T_1) measurements of 2,6-di-*tert*-butylnaphthalene. *Chem. Comm.*, 2000, 651-652.

[40] Fukunaga, T; Kumagae, N; Ishida, H. Crystal structure and cationic motion of *o*-toludinium chloranilate and *m*-toludinium chloranilate studied by X-ray diffraction and H-1 NMR. *Zeitschrift für Naturforschung A*, 2003, 58*a*, 631-637.

[41] Penner, GH; Polson, JM; Stuart, C; Ferguson, G; Kaitner, B. Deuterium NMR and X-ray crystallographic studies of guest and host motions in the thiourea/ 1,4-di-*tert*-butylbenzene inclusion compound. *J. Phys. Chem.*, 1992, 96, 5121-5129.

[42] Sidhu, PS; Penner, GH; Jeffrey, KR; Zhao, B; Wang, ZL; Goh, I. Solid-state NMR study of guest molecule dynamics in 4-alkyl-*tert*-butylbenzene/ thiourea inclusion compounds. *J. Phys. Chem. B*, 1997, 101, 9087-9097.

[43] Riddell, FG; Rogerson, M. The double *tert*-butyl rotation in solid *cis*-2-*tert*-butyl-5-(*tert*-butylsulphonyl)-1,3-dioxane. *Magn. Reson. Chem.*, 1997, 35, 333-336.

[44] Pacansky, J; Yoshimine, M. Theoretical studies on the barriers for internal rotation of the methyl groups in the *tert*-butyl radical. *J. Phys. Chem.*, 1986, 90, 1980-1983.

[45] Richardson, PR; Chapman, MA; Wilson, DC; Bates, SP; Jones, AC. The nature of conformational preference in a number of *p*-alkylphenols and *p*-alkyl benzenes. *Phys. Chem. Chem. Phys.*, 2002, 4, 4910-4915.

[46] Stevenson, PE; Bhat, G; Bushweller, CH; Anderson, WG. Experimental and theoretical studies of the rate of *tert*-butyl rotation in 2-*tert*-butyl-1,3-diheteroatomic rings. The gem-dialkyl effect, ring geometry, and lone pair orientation as factors in conformational dynamics. *J. Am. Chem. Soc.*, 1974, 96, 1067-1071.

[47] Mate, B; Suenram, RD; Lugez, C. Microwave studies of three alkybenzenes: Ethyl, *n*-propyl, and *n*-butylbenzene. *J. Chem. Phys.*, 2000, 113 (1), 192-199.

[48] Speakman, LD; Papas, BN; Woodcock, HL; Schaefer, HFI. The microwave and infrared spectroscopy of benzaldehyde: conflict between theory and experimental deductions. *J. Chem. Phys.*, 2004, 120, (9), 4247-4250.

[49] Moazzen-Ahmadi, N; Kelly, E; Schroderus, J; Horneman, VM. The high-resolution far-infrared torsional spectrum of ethane. *J. Mol. Spectr.*, 2001, 209, 228-232.

[50] Bernstein, ER; Im, HS; Yong, MA. Supersonic jet studies of ethoxybenzenes: geometry of their minimum energy conformations. *J. Org. Chem.*, 1991, 56, 6059-6063.

[51] Breen, PJ; Bernstein, ER; I, SJ; Secor, HV. Determination of the minimum-energy conformation of arylbenzene and its clusters with methane, ethane, water, and ammonia. *J. Phys. Chem.*, 1989, 93, 6731-6736.

[52] Breen, PJ; Warren, JA; Bernstein, ER. Torsional motion in aromatic molecules. Conformational analysis of methyl-, ethyl-, and *n*-propylbenzenes. *J. Am. Chem. Soc.*, 1987, 109, 3453-3455.

[53] Disselkamp, R; Bernstein, ER; I, SJ; Secor, HV. Minimum energy conformation of *ortho*-xylene in its ground and first excited electronic states. *J. Chem. Phys.*, 1992, 97, 8130-8136.

[54] Grassian, VH; Bernstein, ER. Spectroscopy and structure of jet-cooled α-methylstyrene. *J. Phys. Chem.*, 1990, 94, 6691-6695.

[55] Im, HS; Bernstein, ER; Secor, HV; I, SJ. Supersonic jet studies of benzyl alcohols: minimum energy conformations and torsional motion. *J. Am. Chem. Soc.*, 1991, 113, 4422-4431.

[56] Seeman, JI; Paine, JBI; Secor, HV; Im, HS; Bernstein, ER. Supersonic jet studies of alkyl-substituted pyrazines and pyridines. minimum energy conformations and torsional motion. *J. Am. Chem. Soc.*, 1992, 114, 5269-5280.

[57] Pradhan, B; Singh, BP; Nandi, CK; Chakraborty, T; Kundu, T. Origin of methyl torsional barrier in 1-methyl-2-1H-pyridone. *J. Chem. Phys.*, 2005, 122, 204323-204331.

[58] Berg, U; Liljefors, T; Roussel, C; Sandstrom, J. *Acc. Chem. Res.*, 1985, 18, 80.

[59] Jackman, LM; Cotton, FA. *Dynamic NMR spectroscopy.* Academic Press: New York, 1975.

[60] Lomas, JS; Adenier, A. Rate and equilibrium effects of *tert*-alkyl groups on rotamerizationin 3,4-alkylenedioxy-, 3-alkoxy- and 3,4-dialkoxy-2-thienyl(di-*tert*-alkyl)methanols: an IR, NMR, kinetics and MM study. *J. Chem. Soc., Perkin* 2, 2002, 71264-71270.

[61] Oki, M. *Applications of dynamic NMR spectroscopy to organic chemistry.* VCH Publishers: Deerfield Beach, FL, 2005.

[62] Yamamoto, G; Oki, M. *Tetrahedron Lett.*, 1986, 49.

[63] Yamamoto, G; Oki, M. *Bull. Chem. Soc.*, Jpn., 1986, 59, 3597.

[64] Abragam, A. *The principles of nuclear magnetism.* Oxford University Press: Oxford, 1961.

[65] Slichter, CP. *Principles of magnetic resonances.* 3rd ed.; Springer: Berlin, 1990.

[66] Trueblood, KN; Dunitz, JD. Internal molecular motions in crystals. The estimation of force constants, frequencies and barriers from diffraction data. A feasibility study. *Acta Crystallogr. B*, 1983, *B*39, (1), 120-133.

[67] Schiebel, P; Amoretti, G; Ferrero, C; Paci, B; Prager, M; Caciuffo, R. Coherent internal and centre of mass quantum rotation of methyl groups in the *p-tert*-butyl-calix[4]arene(2:1) *p*-xylene supramolecular complex. *J. Phys.: Condens. Matter*, 1998, 10, 2221-2231.

[68] Dovesi, R; Civalleri, B; Orlando, R; Roetti, C; Saunders, VR. *Ab Initio* Quantum Simulation in Solid State Chemistry. In *Reviews in Computational Chemistry*, Lipkowitz, K. B., Larter, R. and Cundari, T. R., Eds. Wiley-VCH, John Wiley & Sons, Inc.: 2005, Vol. 21, 1-125.

[69] Beckmann, PA; Rosenberg, J; Nordstrom, K; Mallory, CW; Mallory, FB. CF_3 rotation in 3-(trifluoromethyl)phenanthrene: Solid state F-19 and H-1 NMR relaxation and Bloch-Wangsness-Redfield theory. *J. Phys. Chem. A*, 2006, 110(11), 3947-3953.

[70] Wang, X; Mallory, FB; Mallory, CW; Beckmann, PA; Rheingold, AL; Francl, MM. CF_3 rotation in 3-(trifluoromethyl)phenanthrene: X-ray diffraction and ab initio electronic structure. *J. Phys. Chem. A*, 2006, 110(11), 3954 - 3960.

Chapter 2

INTERNAL ROTATION IN ISOLATED MOLECULES: AROMATIC HYDROCARBONS

2.1. ABSTRACT

The internal rotations of the *t*-butyl groups and their constituent methyl groups are studied in the isolated molecules of three aromatic hydrocarbons: 3-*t*-butyl-chrysene, 2,6-di-*t*-butylnaphthalene and 1,4-di-*t*-butylbenzene. The potential energy surfaces for the internal rotations, calculated at the HF/6-311+G(d,p)// HF/6-31G(d) level, are presented. The minimum energy paths are identified on the two-dimensional potential energy surfaces to investigate the coupling relationship between the internal rotations of the methyl and *t*-butyl groups. The barriers to the internal rotation of the *t*-butyl groups are 1.14, 1.30 and 0.68 kcal/mol in 3-*t*-butyl-chrysene, 2,6-di-*t*-butylnaphthalene and 1,4-di-*t*-butylbenzene, respectively. The barriers to the internal rotations of the out-of-plane methyl groups are found to be very similar in the three compounds, about 4.00 kcal/mol. The barriers to the relaxed rotation of the in-plane methyl groups are 5.08, 5.38 and 4.17 kcal/mol in 3-*t*-butylchrysene, 2,6-di-*t*-butylnaphthalene and 1,4-di-*t*-butylbenzene, respectively. The rotation of the *t*-butyl groups is accompanied by small variations of the orientation of the methyl groups. The internal rotation of the out-of-plane methyl groups is also coupled to a small degree with the reorientation of the *t*-butyl groups. The in-plane methyl rotation is found to occur simultaneously with the rotation of the *t*-butyl group.

2.2. INTRODUCTION

The theme of this chapter is the internal rotation of *t*-butyl and methyl groups on aromatic systems in the isolated molecules of three representative compounds: 3-*t*-butylchrysene (**3-TC**), 2,6-di-*t*-butylnaphthalene (**2,6-TTN**) and 1,4-di-*t*-butylbenzene (**1,4-TTB**). Experimental measurements of the barriers to the internal rotations have been made in the polycrystalline solids of the compounds using the solid state nuclear magnetic resonance (NMR) relaxometry method [1-5]. The barriers range from 3.3 to 5.7 kcal/mol in the polycrystalline solids of the three compounds. In the order of **3-TC**, **1,4-TTB** and **2,6-TTN**, the barriers to the synchronized reorientation of the overall *t*-butyl group and its in-plane methyl group decrease from 5.7 to 4.3 kcal/mol, while the barriers to the reorientation of two out-of-plane methyl groups increase from 3.3 to 4.3 kcal/mol. It is of interest to learn how the *inter*molecular and *intra*molecular interactions contribute to the barriers and how the molecular structure and crystal packing pattern affect the rotation dynamics. Quantum mechanical electronic structure calculations proved a helpful tool to provide quantitative answers to these questions.

Penner et al. [6-7] used solid state NMR spectra and relaxometry to study the guest and host dynamics of solid 4-alkyl-*t*-butylbenzene/thiourea (and other host molecules) inclusion compounds. The *t*-butyl group rotation activation barrier was found about 2.6 kcal/mol and the methyl rotation barrier was around 2.9 to 3.1 kcal/mol in the channels formed by thiourea. The *t*-butyl rotational dynamics was also studied both in the solution phase and in the solid state with NMR methods for alkane and alkylbenzene compounds [8-16]. Microwave spectroscopy was used to study the *t*-butyl group in the gas phase of *t*-butylbenzene [17-18]. Supersonic jet cooled laser spectroscopy is used to identify the preferred conformation of the *t*-butyl groups in the ground and electronic excited states for the isolated molecules or clusters of *t*-butylbenzene or other compounds [19]. There have not been many computational studies on the internal rotation of *t*-butyl groups [20-24]. Pacansky et al. [24] used restricted open shell Hartree-Fock calculations to study the methyl group rotation in the *t*-butyl radical. Richardson et al. [23] calculated the *t*-butyl group rotational barrier in the isolated molecules of *t*-butylbenzene and 4-*t*-butylphenol at the MP2/6-31G(d) level. The barrier to the *t*-butyl rotation in *t*-butylbenzene was found about 0.5 kcal/mol. Semi-empirical and molecular mechanics methods were used in earlier studies. There needs to be a systematic theoretical study on the problem using *ab initio* calculations.

Methyl groups directly attached to aromatic systems have been studied extensively for their internal rotations using both theoretical and experimental methods [25-36]. The barrier to the internal rotation of these methyl groups in the isolated molecules is found to be proportional to the bond order difference between the two aromatic ring bonds adjacent to the methyl group [33]. Compared with methyl groups, *t*-butyl groups are much bulkier. Does this conclusion hold for *t*-butyl groups as well? Is there a significant steric effect involved due to the size of the *t*-butyl groups? In this chapter, we aim to present the potential energy surface for the internal rotation of the *t*-butyl groups in the isolated molecules, to explore the possibility of coupled rotation between the *t*-butyl groups and the methyl groups, to interpret the intramolecular origin of the barriers, and to discuss how the molecular structure affects the internal rotation of the *t*-butyl groups. The internal rotations of these *t*-butyl and methyl groups in the crystal packing environments will be discussed in the following chapter.

Before discussing the potential energy surface for the internal rotation of *t*-butyl groups, we briefly review the internal rotation dynamics model of *t*-butyl groups. There are four internal rotations in a *t*-butyl group on an aromatic system if the orientation of the aromatic ring is considered being fixed. One is the overall rotation of the whole *t*-butyl group and the other three are the internal rotations of the three constituent methyl groups of the *t*-butyl group. The potential energy operator of the Hamiltonian describing the internal rotation dynamics of a *t*-butyl group is thus written as a function of four rotational coordinates, $V(\tau_0, \tau_1, \tau_2, \tau_3)$, where τ_0 is the internal rotation coordinate of the *t*-butyl group and τ_1, τ_2 and τ_3 for the methyl groups. If it is assumed that the internal rotations of the three methyl groups are independent, the potential energy operator is then simplified to a sum of three components, $V(\tau_0, \tau_1, \tau_2, \tau_3) = V_1(\tau_0, \tau_1) + V_2(\tau_0, \tau_2) + V_3(\tau_0, \tau_3)$. Furthermore, for simplicity at the outset, the *t*-butyl group and methyl groups can be presumed rigid in solving the internal rotation dynamics problem, although small changes of bond angles and bond lengths have been found important to relieve the barriers. The functions V_1, V_2 and V_3 assume the same function form and the potential energy operator is then written as $V(\tau_0, \tau_1, \tau_2, \tau_3) = V_1(\tau_0, \tau_1) + V_1(\tau_0 + 2\pi/3, \tau_2) + V_1(\tau_0 + 4\pi/3, \tau_3)$. The full information on the rotational dynamics of a *t*-butyl group is available given the two-variable potential energy surface function, $V_1(\tau_0, \tau_1)$.

Experimentally, it is found in the crystalline state that the in-plane methyl group and the *t*-butyl group are characterized by the same hopping rate and the out-of-methyl groups are independent. This could be interpreted that the in-plane methyl rotation is coupled with the *t*-butyl rotation. How might this coupling motion happen if indeed it does happen? The calculation of the two-dimensional potential energy surface of $V_1(\tau_0, \tau_1)$ will provide insight into this problem. In this chapter, the potential energy surface (PES) is calculated for each of the isolated molecules of the compounds, which provides information on the contributions from the intramolecular interactions. The results should be compared with the experimental results from the gas phase. (Such experimental results are not available or limited at present.) The PES in the crystalline solids is studied in the cluster models in the following chapter.

We further decompose the potential energy, $V_1(\tau_0, \tau_1)$, into three terms, $V_1(\tau_0, \tau_1) = V_t(\tau_0) + V_m(\tau_1) + V_c(\tau_0, \tau_1)$. The first two terms describe the 0th order potential energy operator for the internal rotation of the overall *t*-butyl group and the methyl group. The third term, $V_c(\tau_0, \tau_1)$, is the coupling potential energy for the two motions. In this chapter, the potential energy surfaces, $V_t(\tau_0)$, $V_m(\tau_1)$ are first presented, followed by the discussion on the two-dimensional potential energy surface, $V_1(\tau_0, \tau_1)$.

In the above internal rotation dynamics model, the internal rotational degrees of freedom are isolated from other motions of the *t*-butyl group, such as bond stretching and bond angle variations, and also from the motions of the other nuclei, (*e.g.* the aromatic ring). To compute the PES, there are two models frequently used to treat the relationship between internal rotation and the other nuclear degrees of freedom [37-39]. One is the so-called 'sudden' approximation, which assumes that the internal rotation is the fastest motion. The other is the so-called 'adiabatic' approximation, which assumes the internal rotation is slow relative to other motions. The PES in the first model is calculated with all the nuclei fixed at their thermal equilibrium positions except for the internal rotation coordinates. Such a rotational process is called a 'rigid' rotation in this thesis. In fact, for the most molecular systems studied here, the internal rotation (in the scale of 10^7 Hz) is a much slower process than other vibration modes (in the scale of 10^{12} Hz). The adiabatic PES is more accurate and this has been supported by many theoretical and experimental studies. In the adiabatic approximation, the minimum energy path is calculated for the internal rotation process. In the results section, we present the PESs for the internal rotation motions of the *t*-butyl groups and methyl groups. In the discussion section, the origin of the rotational barriers and the coupled motions are discussed.

2.3. METHODS

All the quantum mechanics electronic structure calculations were carried out using the *Gaussian03* software package [40] running under Redhat Linux OS on dual-processor Dell Precision 650 workstations. Initial geometries for the isolated molecules were set up with the aid of *GaussView 3.0* (Gaussian, Inc., Wallingford, CT.). The Natural Bond Orbital (NBO) population analysis was processed with *NBO 3.1* [41] incorporated in *Gaussian03*. Structural optimizations of the isolated molecules were carried out at the theoretical level of HF/6-31G(d)//HF/6-31G(d). The potential energy surfaces were computed at the HF/6-311+G(d,p)//HF/6-31G(d) level. The normal mode analyses, which are used to confirm the energy minimum and maximum points, were done at the HF/6-31G(d)//HF/6-31G(d) level. The NBO analyses were conducted at the HF/6-311+G(d,p)//HF/6-31G(d) level.

The potential energy surfaces were drawn with *Origin 7.0* (OriginLab Co., Northampton, MA.). The Renka-Cline gridding method was used to generate the two-dimensional potential energy surfaces. Molecular structures were drawn with *ChemDraw 7.0* and *Chem3D 7.0* (CambridgeSoft, Cambridge, MA.). Other graphs were created with the aid of *Mathematica 5.0* (Wolfram Research, Inc., Champaign, IL.).

2.4. RESULTS

2.4.1. Labeling Scheme

For the convenience of describing the internal rotation coordinates, the benzene, naphthalene, and chrysene rings of the compounds are numbered according to the IUPAC nomenclature rules for their carbon skeleton. See Scheme 2.1. The central quaternary carbons of *t*-butyl groups are labeled with the letter ‘t’, and the carbon atoms of the constituent methyl groups are labeled with the letter ‘m’, and the aromatic ring carbons to which the *t*-butyl groups are attached are labeled with the letter ‘a’. In the molecules of **1,4-TTB** and **2,6-TTN**, the two *t*-butyl groups are differentiated by their positions on the ring, so are the ‘t’ carbon atoms. The three methyl groups of a *t*-butyl group are distinguished by the dihedrals of their Cm-Ct bonds relative to the aromatic ring, such as in-plane and out-of-plane methyl groups. Hydrogen atoms can be referred by the carbon atoms they are bonded to.

The coordinate of the overall rotation of a *t*-butyl group, τ(*t*-butyl), can be represented by one of the three dihedrals formed by the Cm-Ct bonds with the plane of the aromatic ring. Specifically, the following definitions are adopted, τ(CmCtC3C4) for τ(3-*t*-butyl) in **3-TC**, τ(CmCtC2C1) for τ(2-*t*-butyl) and τ(CmCtC6C5) for τ(6-*t*-butyl) in **2,6-TTN**, and τ(CmCtC1C2) for τ(1-*t*-butyl) and τ(CmCtC4C5) for τ(4-*t*-butyl) in **1,4-TTB**. The 0° conformations of the *t*-butyl groups in the three molecules, τ(*t*-butyl) = 0°, are drawn in Scheme 2.1. The coordinates for the internal rotation of methyl groups, τ(methyl), are represented by one of the three dihedrals τ(HCmCtCa) involving the methyl group of interest. The τ(methyl) = 0° is an eclipsed conformation with respect to the CmCt bond and one of the Cm-H bonds is coplanar and syn with respect to the Ct-Ca bond. The τ(methyl) = 60° conformation is staggered with respect to the Cm-Ct bond. In Scheme 2.2, the staggered and eclipsed conformations are drawn for the in-plane methyl group. For the internal rotations in this study, no significant bond angle changes of the *t*-butyl and methyl groups from the values in the ground state were found by the calculations. The local three-fold symmetry is conserved for the *t*-butyl group and methyl groups. A rotation of 120° repeats the original conformation.

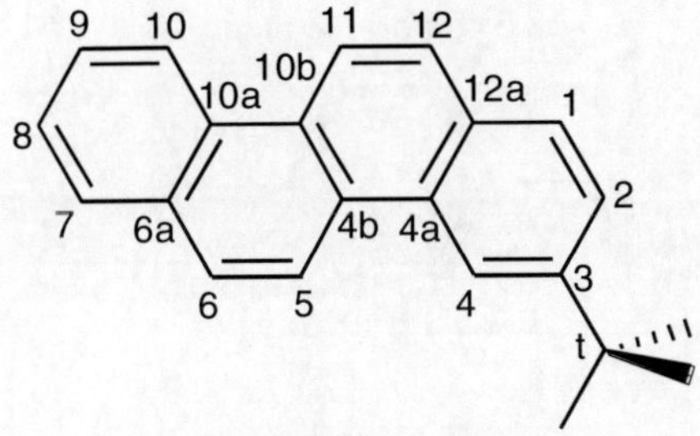

3-*t*-butylchrysene (**3-TC**)

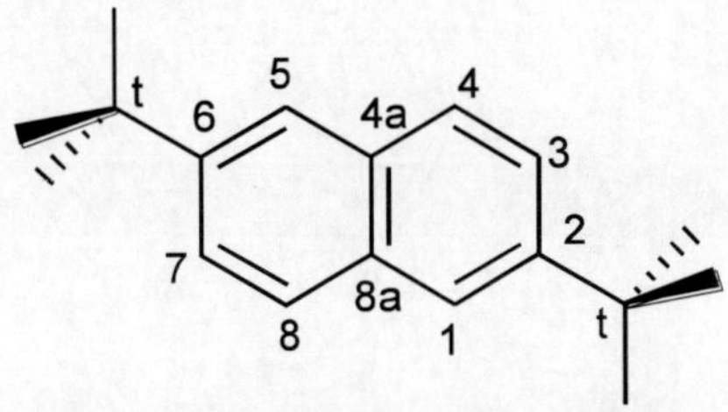

2,6-di-*t*-butylnaphthalene (**2,6-TTN**)

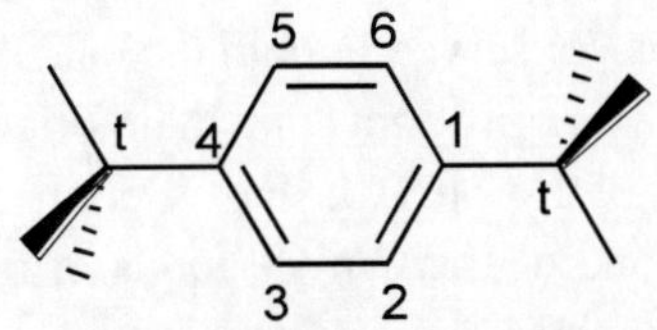

1,4-di-*t*-butylbenzene (**1,4-TTB**)

Scheme 2.1.

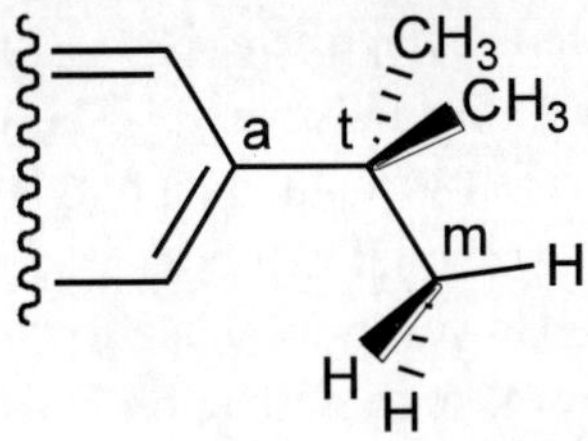

τ(methyl)= 60° (staggered)

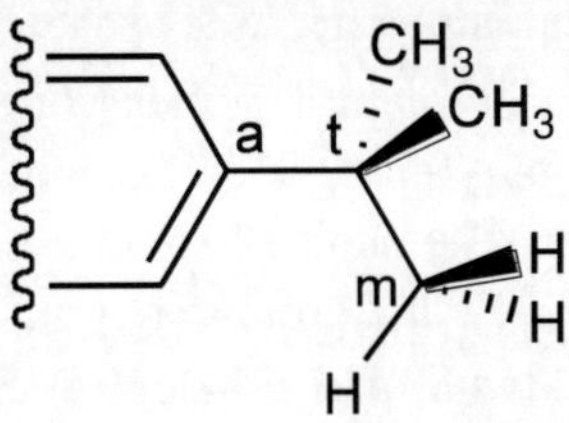

τ(methyl)= 0° (eclipsed)

Scheme 2.2.

2.4.2. Ground State Molecular Structures

The ground state structures of the isolated molecules of **3-TC**, **2,6-TTN** and **1,4-TTB** were obtained with a full structural optimization at the HF/6-31G(d) level. Normal mode analysis of a single molecule at the HF/6-31G(d) level was used to confirm that the optimized structure is the global minimum energy conformation. In general, the computed structures are found to be in good agreement with the crystal structures measured by X-ray diffraction (entry codes in the Cambridge Crystal Database Centre [42] are QEFBAE [43], KOKQUW [1] and BUTBNZ [44], respectively). The structural parameters of the carbon skeleton are compared in Tables 1, 2 and 3 for the three compounds.

For the isolated molecule of **3-TC**, one of the Cm-Ct bonds is coplanar and syn with respect to the ring C3-C4 bond in the computed ground state, *i.e.* τ(3-*t*-butyl) = 0.0°. In the crystal structure, one of the Cm-Ct bonds is nearly coplanar with the ring C3-C4 bond, with τ(3-*t*-butyl) = 1.7°. The carbons in the chrysene ring in the isolated molecule are all in the same plane. In the crystal, the chrysene ring has a slight end-to-end twist of 6.9°. This twisting presumably results from the ring packing interactions in the crystal.

1,4-TTB (computed) **1,4-TTB** (X-ray)

Scheme 2.3.

The molecule of **2,6-TTN** belongs to the C_{2h} point group. The inversion center located halfway along the C9-C10 bond is conserved in both the calculated and crystal structures. The two *t*-butyl groups are chemically equivalent and their orientations are τ(*t*-butyl) = 0°, in which one of the Cm-Ct bonds of 2-*t*-butyl group is coplanar with the ring and syn with respect to the C1-C2 bond, and one of the Cm-Ct bonds of 6-*t*-butyl group is coplanar with the ring and syn with respect to the C6-C5 bond. In the crystal, the thermal equilibrium orientations are nearly the same as in the isolated molecules, τ(*t*-butyl) = 0.3°.

The two *t*-butyl groups in **1,4-TTB** are also equivalent. However, their ground state orientations are found to be different in the computed isolated molecule structure and the X-ray crystal structure. See Scheme 2.3. In the isolated molecule, there is no inversion symmetry, but there is a plane of symmetry instead. The two dihedrals are τ(1-*t*-butyl) = 0° and τ(4-*t*-butyl) = 180°. (The conformation with τ(1-*t*-butyl) = 180° and τ(4-*t*-butyl) = 0° is equivalent. The choice here and below is arbitrary.) The in-plane Ct-Cm bonds on the 1-*t*-butyl and 4-*t*-butyl groups are pointing in the same direction. However, a calculation in which both τ(1-*t*-butyl) and τ(4-*t*-butyl) are forced to be at 0° shows that conformation is only 0.14 kcal/mol higher in energy than the ground state. In the crystal, the inversion center is conserved, but the *t*-butyl groups rotate out from the in-plane position by 8.3°, *i.e.* τ(1-*t*-butyl) = 8.3° and τ(4-*t*-butyl) = -8.3°.

For chrysene and naphthalene, the π-electron density is not uniformly distributed on the rings. The difference is reflected in the C-C bond lengths. Focusing on the local environment for the *t*-butyl group rotation, the C3-C4 bond is shorter than the C3-C2 bond by 0.05 Å in both the calculated and X-ray structures of **3-TC**. See Table 1. The *t*-butyl groups prefer the orientation with the Cm-Ct bond coplanar with the ring and syn with respect to the shorter bond side. This point is similar to that found for the orientation of methyl groups directly attached to aromatic systems, for which it is found the hyperconjugation between the methyl groups with the ring determines the orientation preference with the in-plane C-H bond syn with respect to the shorter ring bond. In **2,6-TTN**, the C2-C1 bond is also found shorter than the C2-C3 bond (and the C6-C5 bond is shorter than the C6-C7 bond) by 0.065 Å in the isolated molecule and 0.054 Å in the crystal structure. The Cm-Ct bonds are coplanar with the ring and syn with respect to the shorter bond side. In **1,4-TTB**, because of the presence of the *t*-butyl groups, the π-electron density on the ring is not uniform as in un-substituted benzene. It is found the C1-C2 bond, which is coplanar and syn with one of the Cm-Ct bonds, is shorter by 0.014 Å than the C1-C6 bond in the isolated molecule. However, the bond length order for molecules is reversed in the crystal for **1,4-TTB**. The C1-C2 bond is longer than the C1-C6 bond by 0.022 Å. The nearly in-plane Cm-Ct bonds are closer to the longer ring bond. Presumably, this results from crystal packing interactions.

The methyl groups are found to be in the staggered conformation in all the isolated molecules, τ(methyl) = 60°. Since hydrogen atoms are difficult to measure in the X-ray diffraction experiments [45-47], the orientation of the methyl groups were not determined. The optimization in the cluster models at the HF/3-21G level was used in Chapter 3 to search the ground state orientation in the crystal. These calculations show that the preferred orientation is the staggered conformation for all the methyl groups in the crystal as well.

Table 2.1 Comparison of the calculated structure with the X-ray diffraction results for 3-*t*-butylchrysene (3-TC)

	HF/6-31G(d)	X-ray
Bond length, Å		
C1C2	1.358	1.365
C2C3	1.419	1.423
C3C4	1.368	1.374
C4C4a	1.420	1.413
C4aC4b	1.453	1.448
C4bC5	1.436	1.433
C5C6	1.344	1.351
C6C6a	1.429	1.419
C6aC7	1.412	1.412
C7C8	1.362	1.364
C8C9	1.406	1.406
C9C10	1.364	1.367
C10C10a	1.417	1.422
C10aC10b	1.454	1.450
C10aC6a	1.404	1.425
C10bC11	1.434	1.433
C10bC4b	1.394	1.413
C11C12	1.345	1.354
C12C12a	1.427	1.418
C12aC4a	1.400	1.422
C12aC1	1.413	1.410
C3Ct	1.537	1.540
CtCm, in-plane[a]	1.536	1.544
CtCm, out-of-plane[b]	1.542	1.544
Bond angle,°		
C4C3Ct	123.1	123.7
C3CtCm, in-plane[a]	112.6	112.3
C3CtCm, out-of-plane[b]	109.5(109.5)	109.1(109.0)
Bond dihedral,°		
C4C3CtCm, in-plane[a]	0.0	1.7
C4C3CtCm, out-of-plane[b]	-120.0(120.0)	-118.7(121.6)

[a] In-plane or nearly in-plane methyl carbon.

[b] Two out-of-plane methyl carbons, one in parenthesis.

Table 2.2. Comparison of the calculated structure with the X-ray diffraction results for 2,6-di-*t*-butylnaphthalene (2,6-TTN)

Parameter	HF/6-31G(d)	X-ray
Bond length, Å		
C1C2	1.363	1.369
C2C3	1.428	1.423
C3C4	1.356	1.363
C4C4a	1.419	1.414
C4aC5	1.421	1.419
C5C6	1.363	1.369
C6C7	1.428	1.423
C7C8	1.356	1.363
C8C8a	1.421	1.414
C8aC4a	1.419	1.412
C8aC1	1.401	1.419
C2Ct	1.536	1.533
C6Ct	1.536	1.533
CtCm, in-plane[a]	1.536	1.527
CtCm, out-of-plane[b]	1.541	1.532
Bond angle,°		
α(C2CtCm), in-plane[a]	112.4	111.9
α(C2CtCm), out-of-plane[b]	109.6(109.6)	109.2(109.3)
Bond dihedral,°		
τ(C1C2CtCm), in-plane[a]	0.0	0.3
τ(C1C2CtCm), out-of-plane[b]	-120.0(120.0)	-119.1(121.1)

[a] In-plane or nearly in-plane methyl carbon.
[b] Two out-of-plane methyl carbons, one in parenthesis.

Table 2.3. Comparison of the calculated structure with the X-ray diffraction results for 1,4-di-*t*-butylbenzene(1,4-TTB)

Parameter	HF/6-31G(d)	X-ray
Bond Length, Å		
C1C2	1.384	1.396
C2C3	1.392	1.401
C3C4	1.384	1.374
C4C5	1.398	1.396
C5C6	1.378	1.401
C6C1	1.398	1.374
C1Ct	1.537	1.537
CtCm, in-plane[a]	1.536	1.530
CtCm, out-of-plane[b]	1.541(1.541)	1.528(1.537)
Bond angle,°		
α(C1CtCm), in-plane[a]	112.3	111.9

Parameter	HF/6-31G(d)	X-ray
α (C1CtCm), out-of-plane[b]	109.6(109.6)	108.3(110.0)
Bond dihedral,°		
τ(C2C1CtCm), in-plane[a]	0.0	8.3
τ (C2C1CtCm), out-of-plane[b]	-120.0(120.0)	-111.9(128.4)

[a] In-plane or nearly in-plane methyl carbon.
[b] Two out-of-plane methyl carbons, one put in parenthesis.

Table 2.4. The least-square best fit results of the three-fold (V_3) and six fold (V_6) components of the *t*-butyl rotation potentials in the isolated molecules

Compound	V_3, kcal/mol	V_6, kcal/mol
3-*t*-butylchrysene	0.96 ± 0.03	0.58 ± 0.03
2,6-di-*t*-butylnaphthalene	1.17 ± 0.03	0.56 ± 0.03
1,4-di-*t*-butylbenzene	0.12 ± 0.02	0.63 ± 0.02
t-butylbenzene	0.02 ± 0.02	0.62 ± 0.02

2.4.3. Potential Energy Surfaces for the Internal Rotation of the *t*-Butyl Groups

The fully relaxed potential energy surfaces with respect to the internal rotation of *t*-butyl groups in the isolated molecules were calculated at the HF/6-311+G(d,p) //HF/6-31G(d) level. They are shown in Figure 2.1, along with the PES for the internal rotations of *t*-butyl groups in the isolated molecule of *t*-butylbenzene (**TB**)*. Only the region from 0 to 60° is shown. At other rotational dihedrals, the PES can be easily obtained by the symmetry properties. The calculations were carried out by fixing the specific τ(*t*-butyl) dihedral in the range from 0 to 60° in increments of 15° and fully optimizing all the other structural parameters at the HF/6-31G(d) level. Only one *t*-butyl group of the two *t*-butyl groups in **1,4-TTB** and **2,6-TTN** is rotated here and the other one is free to adopt its energy minimum conformation.

As seen in Figure 2.1, the τ(*t*-butyl) = 0° conformation is the ground state for all the compounds. The τ(*t*-butyl) = 60° is an intermediate state, *i.e.* a higher-energy local minimum energy point for the *t*-butyl group rotation in **3-TC**, **2,6-TTN** and **1,4-TTB**. The PES for **TB** has 6-fold symmetry, *i.e.* the conformation of τ(*t*-butyl) = 60° is equivalent to the conformation of τ(*t*-butyl) = 0°. The rotational transition state is computed to have τ(*t*-butyl) = 45° conformation in the isolated molecules of **3-TC** and **2,6-TTN**. In the isolated molecule of **1,4-TTB**, the computed transition state has τ(*t*-butyl) = 30°.

The rotational barriers, defined as the energy difference between the transition and ground states, are 0.63 kcal/mol in the isolated molecule of **TB**, 0.68 kcal/mol in the isolated molecule of **1,4-TTB**, 1.30 kcal/mol in the isolated molecule of **2,6-TTN** and 1.14 kcal/mol in the isolated molecule of **3-TC**. The energy differences between the intermediate state and the ground state is 0.14, 1.14 and 0.93 kcal/mol for the isolated molecules of **1,4-TTB**, **2,6-**

* An additional set of results for the *t*-butyl rotation in **TB** with the benzene ring carbon skeleton being fixed at their positions in the ground state is also shown. The results are nearly identical to the fully relaxed rotation and this conclusion will be used in the following chapter.

TTN and **3-TC**, respectively. The calculated barrier of 0.63 kcal/mol for the isolated molecule of **TB** is in good agreement with the value of 0.62 kcal/mol obtained by Richardson et al. at the MP2/6-31G(d) level [12].

The PES curves can be fitted to a sum of three-fold cosine and six-fold cosine potential energy functions, $V(\tau)=\frac{V_3}{2}(1-\cos 3\tau)+\frac{V_6}{2}(1-\cos 6\tau)$.(†) The fitted three-fold components (V_3) and six-fold components (V_6) are listed in Table 4. The fitted curves are plotted in solid lines in Figure 2.1. Seen from Table 4, the three-fold components increase from 0 to 1.17 kcal/mol in the order of **TB**, **1,4-TTB**, **3-TC** and **2,6-TTN**, while the six-fold components stay nearly the same in the region from 0.56 to 0.63 kcal/mol.

During the *t*-butyl group rotation process in the isolated molecules, the methyl groups are found to adjust their orientations slightly. In the range from τ(*t*-butyl) = 0 to 30°, the τ(methyl) of the in-plane methyl group (labeled by τ(*t*-butyl)) and one of the out-of-plane methyl groups (labeled by τ(*t*-butyl) +120°) change from 60° to 55°. From the range τ(*t*-butyl) = 30 to 60°, the reverse changes happen and the methyl groups each go back to the staggered conformation, τ(methyl) = 60°. On further rotation of the *t*-butyl group from 60 to 120°, the same variations are expected to happen for the two methyl groups labeled by the dihedrals of τ(*t*-butyl) +120° and τ(*t*-butyl) -120°.

2.4.4. Potential Energy Surfaces for the Internal Rotation of the Constituent Methyl Groups

As seen above, the *t*-butyl rotational ground state is the τ(*t*-butyl) = 0° conformation for the four compounds. In the ground state, the three constituent methyl groups of a *t*-butyl group can be classified into two types according to their positions relative to the aromatic ring. One is the in-plane methyl group with the Cm-Ct bond coplanar with the ring and the other two are the out-of-plane methyl groups with the Cm-Ct bond forming a dihedral of 120° (or 60°) with the ring plane. The methyl groups are identified by the corresponding τ(*t*-butyl) dihedrals. The in-plane methyl group is with its τ(*t*-butyl) = 0° and the out-of-plane methyl groups are with a τ(*t*-butyl) of +120 or -120°, approximately.

The rotational PESs for the two types of methyl groups were calculated in a similar way as for *t*-butyl rotation from 0 to 60° at the HF/6-311G(d,p)//HF/6-31G(d) level. The PESs for the internal rotation of the out-of-plane methyl groups in the isolated molecules are shown in Figure 2.2. The PESs for the internal rotation of the in-plane methyl groups are shown in Figure 2.3. The structure is fully optimized at the HF/6-31G(d) level without any additional constraint except the coordinates of τ(methyl). The coordinates of τ(*t*-butyl) are also optimized and they are found to reorient accompanying the internal rotation of the methyl groups.

† All the fittings of the potential energy surfaces in this book are linear (with respect to the fitting variables) using the cosine potential functions as the fitting basis. For a typical fit, the regression R^2 value is better than 0.99. All the fitting components along with the barriers are reported up to 0.01 kcal/mol regardless of the fitting uncertainties.

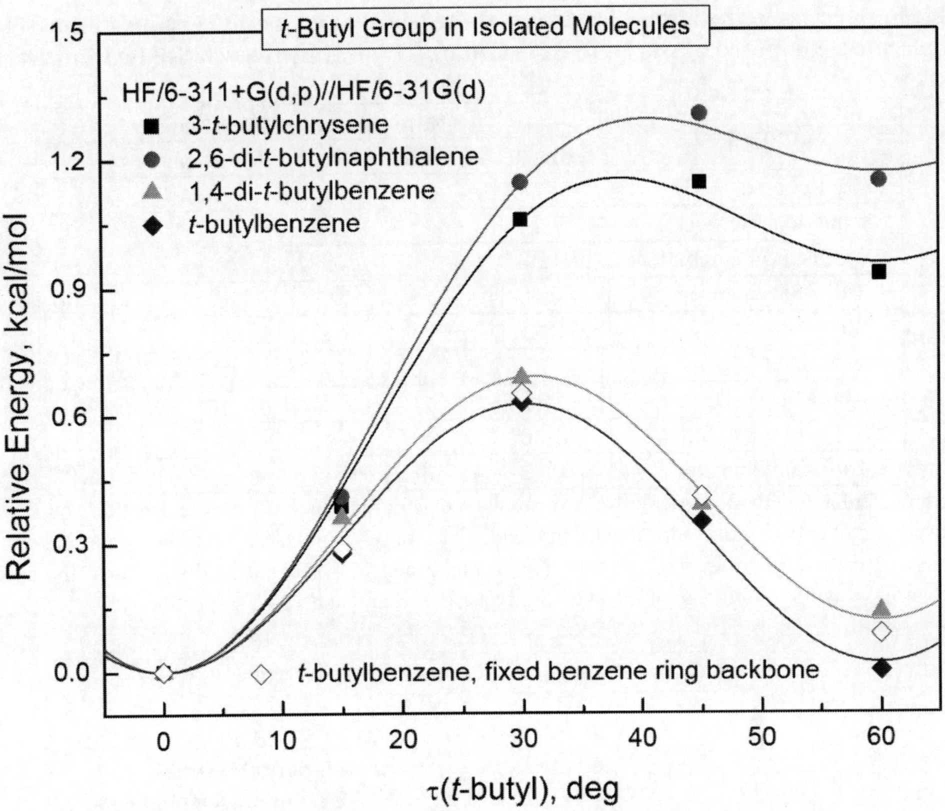

Figure 2.1. The fully relaxed rotational potential energy surfaces for the internal rotation of *t*-butyl groups in the isolated molecules, calculated at the HF/6-311+ G(d,p)//HF/6-31G(d) level. The solid lines are the least-square best fits to the function, $V(\tau)=\frac{V_3}{2}(1-\cos 3\tau)+\frac{V_6}{2}(1-\cos 6\tau)$. The best-fit V_3 and V_6 components are listed in Table 4.

For all the methyl groups, the τ(methyl) = 60° conformation, which is a staggered conformation, is the rotational ground state. The eclipsed conformation, τ(methyl) = 0°, is the rotational transition state. The three curves for the internal rotation of the out-of-plane methyl groups are nearly overlapping. See Figure 2.2. The rotational barriers are 4.00, 3.99 and 3.99 kcal/mol in **1,4-TTB**, **2,6-TTN** and **3-TC**, respectively. For the internal rotation of the in-plane methyl groups in the isolated molecules (Figure 2.3), the three curves follow similar trends qualitatively, but the barriers are different. The barrier is highest in **2,6-TTN**, about 5.38 kcal/mol. In **3-TC**, the barrier is slightly lower, about 5.08 kcal/mol. The barrier decreases to 4.17 kcal/mol in **1,4-TTB**. The barriers for the in-plane methyl group rotations in these three compounds follow the same relative order as the barriers for the *t*-butyl groups.

The PESs for the methyl rotations can be fitted well with a three-fold potential energy function, $V(\tau)=\frac{V_3}{2}(1-\cos 3(\tau-\frac{\pi}{3}))$. (The phase $-\pi/3$ reflects that the ground state is τ(methyl) = 60°.) The fitted curves are plotted in solid lines in Figures 2.2 and 2.3.

Table 5. Barriers to the internal rotation of the *t*-butyl and methyl groups in the isolated molecules, computed at the HF/6-311+G(d,p)//HF/6-31G(d) level. See text for details

Compound	Rotational barrier, kcal/mol		
	t-Butyl	In-plane methyl	Out-of-plane methyl
Fully Relaxed[a]			
3-*t*-butylchrysene	1.14	5.08	3.99
2,6-di-*t*-butylnaphthalene	1.30	5.38	3.99
1,4-di-*t*-butylbenzene	0.68	4.17	4.00
t-butylbenzene	0.63	4.12	4.12
Constrained[b]			
3-*t*-butylchrysene	1.15	6.61	4.19
2,6-di-*t*-butylnaphthalene	1.40	6.42	3.99
1,4-di-*t*-butylbenzene	0.72	6.47	4.17

[a] Barrier calculated with the structural fully optimized except the specified internal rotation coordinate.

[b] Barrier calculated with one additional constraint. For *t*-butyl groups, the rotational coordinate of the related methyl group is fixed at 60°. For in-plane and out-of-plane methyl groups, the specific internal rotation coordinate of the *t*-butyl group is fixed at 0 and 120°, respectively.

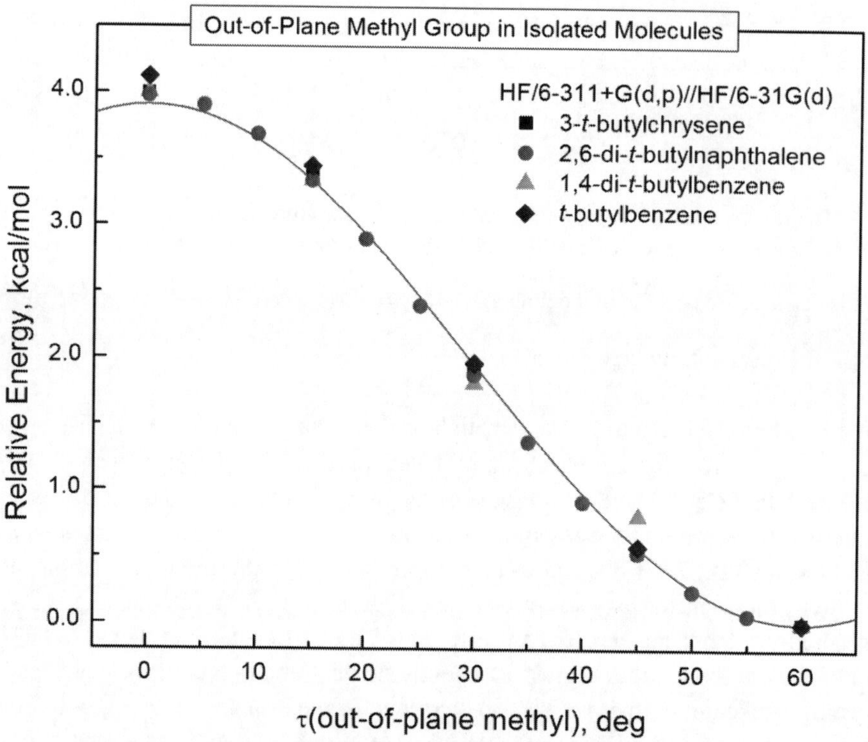

Figure 2.2. The fully relaxed rotational potential energy surfaces of the out-of-plane methyl groups in the isolated molecules, calculated at the HF/6-311+G(d,p) //HF/6-31G(d) level. The solid line is the least-square fit to the function, $V(\tau)=\frac{V_3}{2}(1-\cos 3(\tau-\frac{\pi}{3}))$.

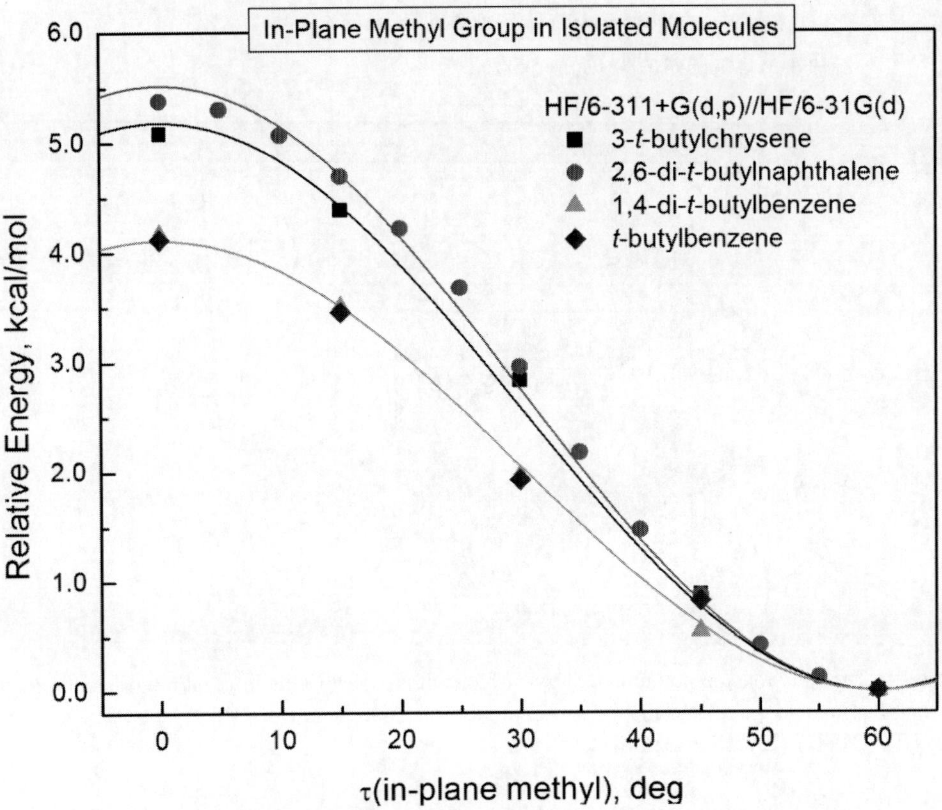

Figure 2.3. The fully relaxed rotational potential energy surfaces of the in-plane methyl groups in the isolated molecules, calculated at the HF/6-311+G(d,p)//HF/6-31G(d) level. The solid lines are the least-square fit to the function, $V(\tau)=\frac{V_3}{2}(1-\cos 3(\tau-\frac{\pi}{3}))$.

During the rotational process of the out-of-plane methyl groups in the isolated molecules, the optimized orientation of the *t*-butyl groups is found to change by only about 5°. For example, if the out-of-plane methyl group of τ(*t*-butyl) = -120° rotates from the ground state (τ(methyl) = 60°) to the transition state (τ(methyl) = 0°), the τ(*t*-butyl) will change to -125°. Similar relaxations are found in all the three compounds. However, there is significant change of the *t*-butyl orientation accompanying the in-plane methyl rotation. At the rotational ground state (τ(methyl) = 60°) of the in-plane methyl group in the isolated molecule of **3-TC**, the optimized τ(*t*-butyl) is found to be 0.4°, which is the rotational ground state for the *t*-butyl group. When rotating the methyl group by 60° to the transition state, τ(methyl) = 0°, the optimized orientation of the *t*-butyl group is found to be τ(*t*-butyl) = 54.6°, which is nearly a *t*-butyl rotation intermediate state. The in-plane methyl group is not 'in-plane' anymore, but has become an out-of-plane methyl in the higher-energy intermediate state with respect to the *t*-butyl rotation. Similar changes are also found for the in-plane methyl rotations in the isolated molecules of **2,6-TTN** and **1,4-TTB**.

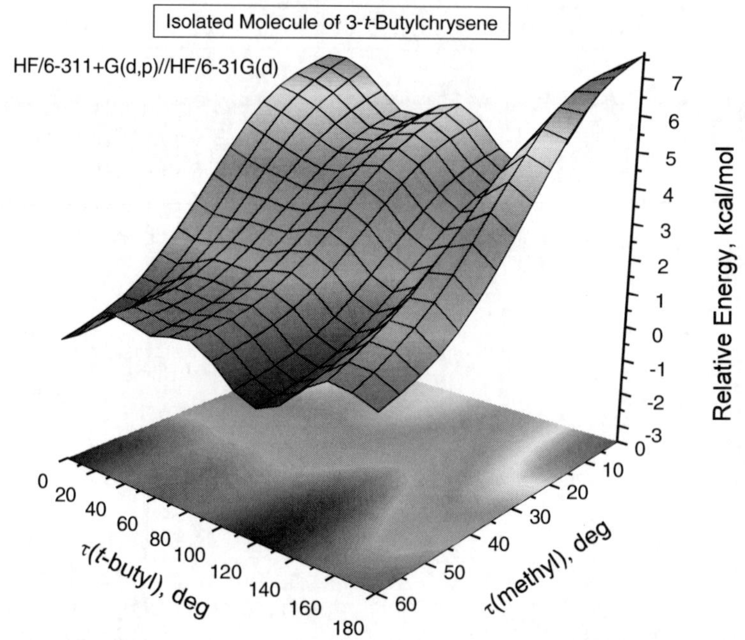

Figure 2.4. The fully relaxed potential energy surface with respect to the internal rotation coordinates of the *t*-butyl group and the related methyl group in the isolated molecule of 3-*t*-butylchrysene, calculated at the HF/6-311+G(d,p)//HF/6-31G(d) level.

2.4.5. Two-Dimensional Potential Energy Surfaces Obtained from Considering the Superimposed Rotation of *t*-Butyl Groups and Their Constituent Methyl Groups

To investigate the coupled relationship between the internal rotations of the *t*-butyl group and the constituent methyl groups, the two-dimensional PES was calculated at the HF/6-311+G(d,p)//HF/6-31G(d) level. The two internal rotation coordinates of τ(*t*-butyl) and the related τ(methyl), in which the two coordinates share a same methyl carbon Cm, are fixed and other structural parameters are allowed to be fully optimized.

A full range calculation was done for the isolated molecule of **3-TC**, as shown in Figure 2.4. The τ(*t*-butyl) coordinate runs from 0 to 180° and the τ(methyl) coordinate runs from 0 to 60°. A complete PES ranging 0 to 360° for both the coordinates can be easily obtained according to the symmetry properties.

The PES in Figure 2.4 can be cut into three regions, τ(*t*-butyl) from 0 to 60°, 60 to 120° and 120 to 180°. In the first range, τ(*t*-butyl) group from 0 to 60°, the related methyl group changes from an in-plane methyl group to an out-of-plane methyl group in the intermediate state of the *t*-butyl rotational state (τ(*t*-butyl)=60°). In the second range of τ(*t*-butyl) = 60 to 120°, the methyl group changes from an out-of-plane methyl group in the *t*-butyl group's intermediate state to an out-of-plane methyl group in the *t*-butyl group's ground state. In the third range, the methyl group converts from an out-of-plane methyl group in the *t*-butyl group's ground state to an in-plane methyl group in the *t*-butyl group's intermediate state. The first and the third ranges contain the most interesting information on the synchronized reorientation of the in-plane methyl and *t*-butyl groups.

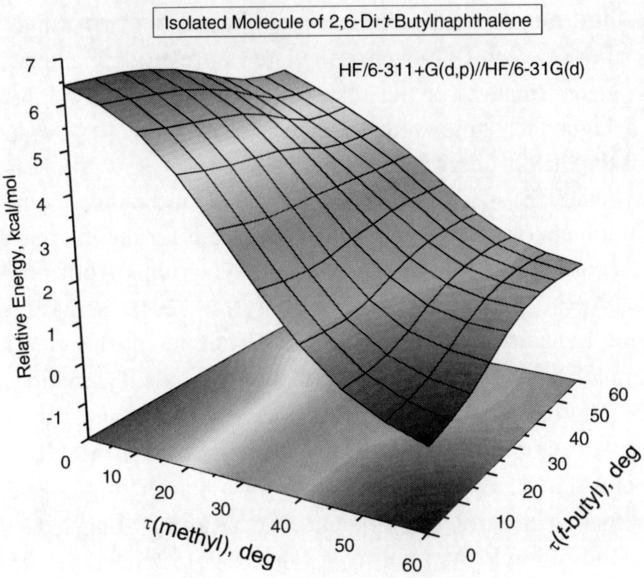

Figure 2.5A.

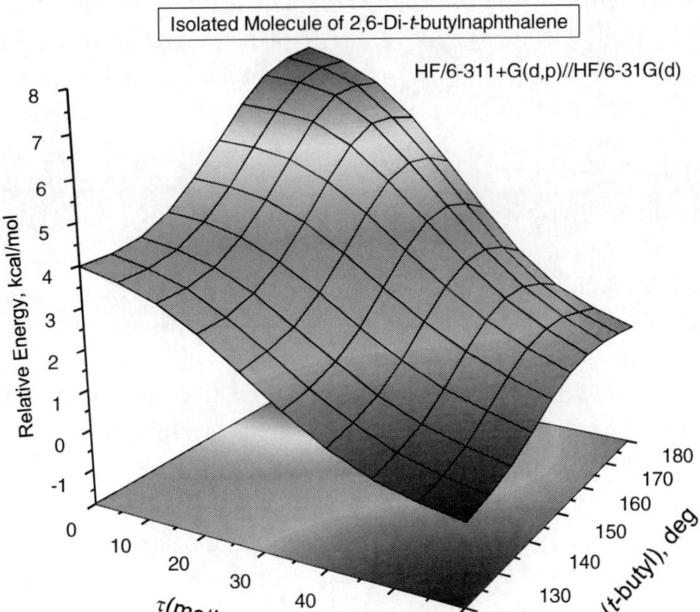

Figure 2.5B. The fully relaxed potential energy surface with respect to the internal rotation coordinates of one of the *t*-butyl groups and the related methyl group in the isolated molecule of 2,6-di-*t*-butylnaphthalene, calculated at the HF/6-311+G(d,p)//HF/6-31G(d) level.

For the isolated molecules of **2,6-TTN**, the calculations were done in the two regions, from τ(*t*-butyl) = 0 to 60° and from 120 to 180°. They are shown in Figure 2.5. For the isolated molecule of **1,4-TTB**, the PES was computed in the range from 0 to 60°. It is shown in Figure 2.6. The third range is expected to be similar as the first range. The PES for the isolated molecules of **3-TC**, **2,6-TTN** were calculated for every 15° interval for both the τ(*t*-butyl) and τ(methyl) coordinates, for the sake of computational cost. The PES of **1,4-TTB** was calculated in a higher resolution, for every 5° interval along both the coordinates. The detailed structure of the PES is observable in this PES.

The two dimensional PESs for the isolated molecules of the three compounds are similar in shape. The minimum energy paths (MEPs) can be found for the internal rotation of *t*-butyl and methyl groups. For the internal rotation of *t*-butyl groups from one global minimum energy point (τ(*t*-butyl) = 0°) to another one (τ(*t*-butyl) = 120°), the MEP passes two saddle points, which are the transition states, and one local minimum, the intermediate state. The MEP is mainly parallel to the τ(*t*-butyl) axis and only a small variation occurs along the τ(methyl) direction. The barrier to cross the transition state is small, about 1.3 kcal/mol or lower. For the internal rotation of an out-of-plane methyl group, the MEP is along a trough parallel to the τ(methyl) axis. A transition state with a barrier of 4.0 kcal/mol needs to be overcome. The internal rotation of the in-plane methyl groups (labeled by τ(*t*-butyl) = 0°) follows a different kind of MEP. As seen from Figures 2.4, 2.5 and 2.6, if a path parallel to the τ(methyl) axis is taken, a barrier as high as 6.5 kcal/mol needs to be passed. To avoid this high barrier, the MEP takes a diagonal route from the global minimum energy point (τ(*t*-butyl) = 0° and τ(methyl) = 60°) to the saddle point (τ(*t*-butyl) = 60° and τ(methyl) = 0°). As seen in the fully relaxed calculations above, the in-plane methyl groups at the ground state convert to the out-of-plane methyl groups in the *t*-butyl rotational intermediate state. By assuming this MEP, the barrier decreases significantly from 6.6 to 5.1 kcal/mol.

More detailed information on the coupling between the *t*-butyl and methyl groups can be found in the high-resolution PES of **1,4-TTB** in Figure 2.6. During the rotational process of the *t*-butyl group, the methyl groups stay at their ground state orientation (τ(methyl) = 60°). However, a small relaxation, τ(methyl) decreasing by 5°, occurs at τ(methyl) = 30°. Such small magnitude relaxation is observed for the out-of-plane methyl groups. The phenomena here are consistent to those found in the fully relaxed one-dimensional rotation calculations.

One way to view the two-dimensional PESs is to take them as the dependence relationship of the constrained methyl rotation on the relative position of the methyl group labeled by τ(*t*-butyl). We call the methyl rotation along the straight path parallel with the τ(methyl) axis as the 'constrained' methyl rotation. In Figure 2.7, the barriers of the constrained methyl rotations are plotted against τ(*t*-butyl). In Table 5, the barriers at the two particular positions (τ(*t*-butyl) = 0 and 120°) are listed as the constrained barrier for the in-plane and out-of-plane methyl rotations. The constrained barrier for the *t*-butyl rotation is also reported in the table, in which the straight path at τ(methyl)= 60° is assumed for the *t*-butyl rotation. If the PES of the constrained methyl rotation is assumed to be a three-fold cosine potential and only the constrained rotation barrier is dependent on the coordinate of τ(*t*-butyl), we can separate the two variables, τ(methyl) and τ(*t*-butyl), of the two-dimensional PESs in the following way,

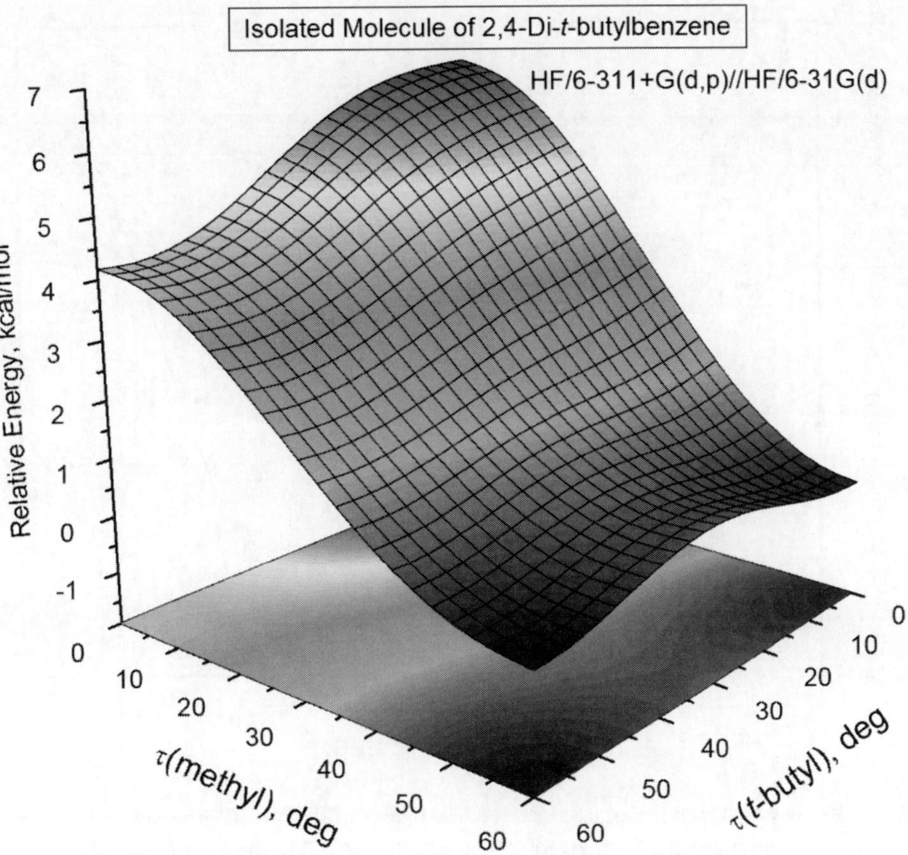

Figure 2.6. The fully relaxed potential energy surface with respect to the internal rotation coordinates of one of the *t*-butyl groups and the related methyl group in the isolated molecule of 1,4-di-*t*-butylbenzene, calculated at the HF/6-311+G(d,p) //HF/6-31G(d) level.

$$V(\tau_t, \tau_m) = V_t(\tau_t) + \frac{V_m(\tau_t)}{2}(1 - \cos 3(\tau_m - \frac{\pi}{3})).$$

The dependence of $V_m(\tau_t)$, the constrained methyl barrier, on τ_t can be obtained from the data in Figure 2.7, which can be fitted to the following equation, $V_m(\tau_t) = V_{m0}1 + V_{m1}\cos\tau_t + V_{m2}\cos 2\tau_t + V_{m4}\cos 4\tau_t$.

The term of $V_t(\tau_t)$ is the *t*-butyl rotational barrier, given in the previous section for the fully relaxation study. The fitted curve of $V_m(\tau_t)$ is plotted with a solid line in Figure 2.7 for the isolated molecule of **3-TC**. As seen in Figure 2.7, the results from the three compounds are nearly identical. The choice of the fitting cosine functions is to retain the symmetry property of the PES. The best fit of the 0th order term, V_{m0}, is 9.57 ± 0.11 kcal/mol and the other coefficients are 0.30 ± 0.15, 1.88 ± 0.15 and 1.59 ± 0.15 kcal/mol for V_{m1}, V_{m2} and V_{m4}, respectively.

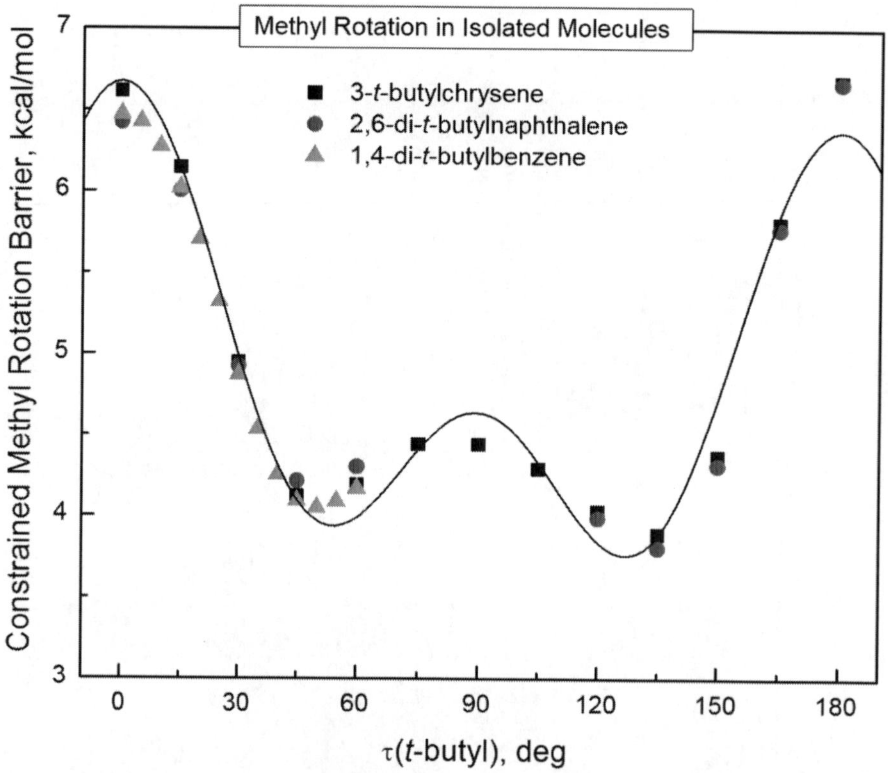

Figure 2.7. The dependence of the constrained rotation barrier of the methyl group on the relative orientation of the *t*-butyl group, τ(*t*-butyl) for all three molecules. The barriers are from the two-dimensional potential energy surfaces in Figures 2.4, 2.5 and 2.6. The solid line is the least-square best fit to the function, $V_m(\tau_t) = V_{m0}1 + V_{m1}\cos\tau_t + V_{m2}\cos 2\tau_t + V_{m4}\cos 4\tau_t$.

2.4.6. Two-Dimensional Potential Energy Surfaces Obtained from Considering the Rotation of Two *t*-Butyl Groups in 2,6-TTN and 1,4-TTB

There are two equivalent *t*-butyl groups in **2,6-TTN** and **1,4-TTB**. Are their internal rotation motions correlated? To answer the problem, the two-dimensional PESs with respect to the rotational coordinates of the two *t*-butyl groups were calculated. They are shown in Figures 2.8 and 2.9. The PESs can be roughly seen as a sum of the potential energy curves of two independent relaxed *t*-butyl rotations. (Mathematically, this means that the PES can be fitted to the equation, $V(\tau_1, \tau_2) = V_t(\tau_1) + V_t(\tau_2)$.) The internal rotations of the *t*-butyl groups are approximately independent from each other, although there is a slight coupling effect, as seen in Tables 2.6 and 2.7, in which the rotational barrier of one of the *t*-butyl groups is tabulated against the orientation of the other *t*-butyl group.

Table 2.6. The dependence of the barrier to the internal rotation of one of the *t*-butyl groups (2-*t*-butyl) on the orientation of the other *t*-butyl group (6-*t*-butyl) in the isolated molecule of 2,6-di-*t*-butylnaphthalene, computed at the HF/6-311+G (d,p)//HF/6-31G(d) level

τ(6-*t*-butyl),°	0	15	30	45	60
Barrier for 2-*t*-butyl, kcal/mol	1.30	1.31	1.33	1.38	1.42

Table 7. The dependence of the barrier to the internal rotation of one of the *t*-butyl groups (1-*t*-butyl) on the orientation of the other *t*-butyl group (4-*t*-butyl) in the isolated molecule of 1,4-di-*t*-butylbenzene, computed at the HF/6-311+G(d,p) //HF/6-31G(d) level

τ(4-*t*-butyl), °	0	15	30	45	60
Barrier for 1-*t*-butyl, kcal/mol	0.69	0.68	0.63	0.66	0.68

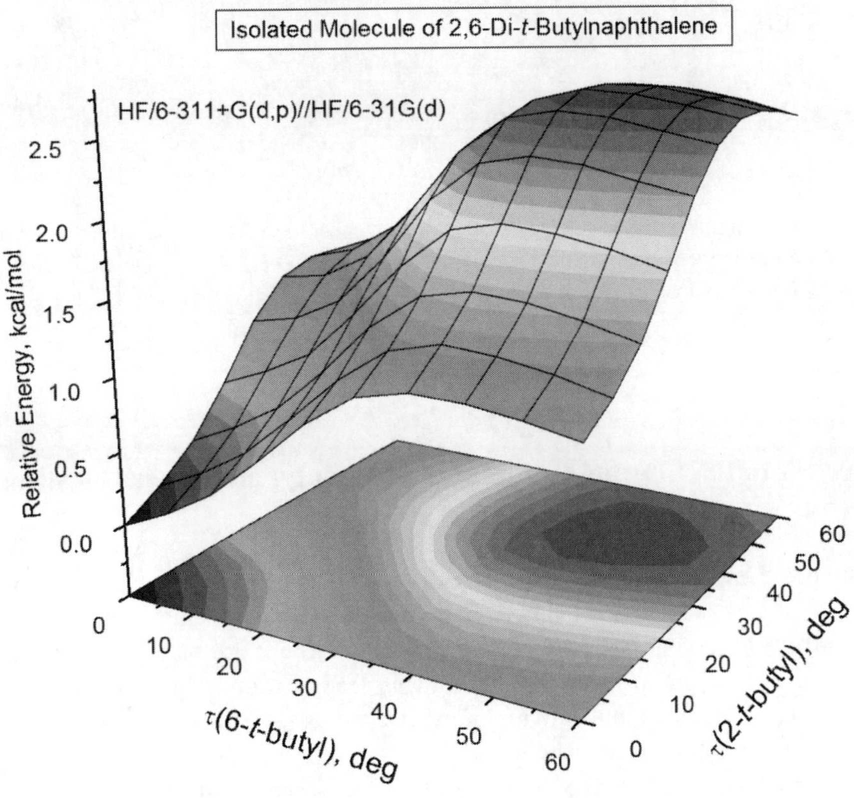

Figure 2.8. The fully relaxed potential energy surface with respect to the internal rotation coordinates of the two *t*-butyl groups in the isolated molecule of 2,6-di-*t*-butylnaphthalene, calculated at the HF/6-311+G(d,p)//HF/6-31G(d) level.

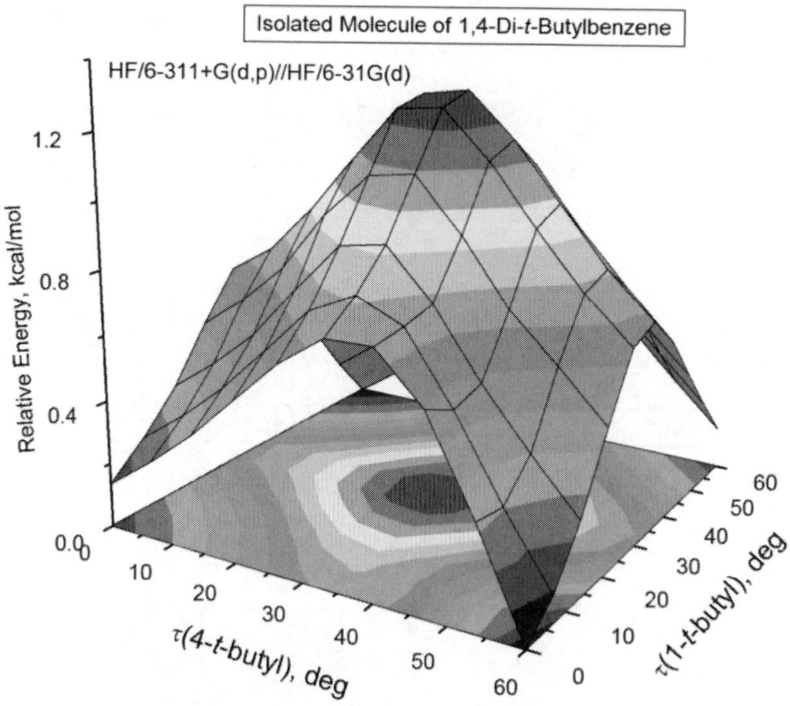

Figure 2.9. The fully relaxed potential energy surface with respect to the internal rotation coordinates of the two *t*-butyl groups in the isolated molecule of 1,4-di-*t*-butylbenzene, calculated at the HF/6-311+G(d,p)//HF/6-31G(d) level.

2.5. Discussion

2.5.1. Barriers to the Internal Rotation of the *t*-Butyl and Their Constituent Methyl Groups

The computed rotational barriers in the isolated molecules are collected in Table 2.5. Compared with the reorientation activation energy measured experimentally in the crystalline solids, the barriers to the internal rotations of the *t*-butyl groups are much smaller while the barriers for the methyl groups are very close. This leads to an early impression that the internal rotations of the *t*-butyl groups in the crystals are strongly affected by the intermolecular interactions in the crystal and the methyl groups are less influenced by the crystal packing. The more detailed study of this idea is the subject of the following chapter.

Among the three compounds, the *t*-butyl rotational barrier is smallest in **1,4-TTB** and largest in **2,6-TTN**. As presented in the results section, the PESs can be decomposed into the three-fold and six-fold components. Their coefficients are reported in Table 4. The six-fold components have little variation across the compounds and the three-fold components have the same trend as the barriers. The three-fold components account for the difference of the barriers among the compounds.

The rotational barriers of the methyl groups are more dependent on the local environment of the *t*-butyl groups, and the size of the aromatic systems has little effect. As shown in Figure

2.7, the constrained rotational barriers of the methyl groups in the three compounds have a similar dependence on the orientation of the *t*-butyl groups, independent of the aromatic ring structure. This is also true for the fully relaxed rotational barriers of the out-of-plane methyl groups as shown in Table 5. However, there are large differences among the three compounds for the fully relaxed rotations of the in-plane methyl groups. As seen in Table 5, the barriers are in the same trend as the barriers of the *t*-butyl groups. The reason is due to the concerted internal rotation of the in-plane methyl group and the *t*-butyl group: the in-plane methyl group at its the ground state changes to the out-of-plane methyl group in the *t*-butyl rotational intermediate state at the methyl rotation transition state. Therefore, the difference between the fully relaxed barriers of the in-plane and out-of-plane methyl groups is determined by the relative energy of the *t*-butyl rotational intermediate state, which is proportional to the *t*-butyl barrier.

2.5.2. Role of Intramolecular Interactions in Rotational Barriers

There are many different models proposed for interpreting the rotational barrier in ethane and other molecules. We discuss the barriers to the internal rotation of the *t*-butyl groups in terms of hyperconjugation and steric effect in a qualitative way. For a more complete quantitative analysis, higher level calculations are necessary to treat the electron correlation effect more accurately. This is beyond the scope of this study.

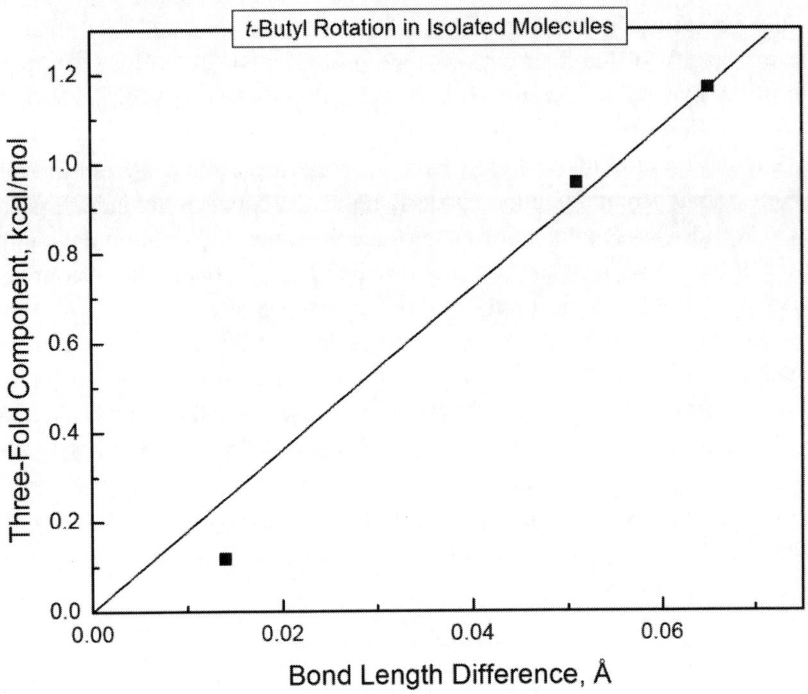

Figure 2.10. The dependence of the three-fold component of the *t*-butyl rotation PESs in the isolated molecule on the bond length difference between the two ring C-C bonds adjacent to the *t*-butyl group. The straight line is the least-square best fit of the data.

Hyperconjugation is found to determine the barrier of the methyl rotation on the aromatic ring [31-36]. The barrier is linearly correlated with the bond order difference adjacent to the methyl group, in absence of steric effects [33]. In toluene, the two aromatic bonds adjacent to the methyl group have the same bond order, and the rotational barrier is very small, only a few calories per mol. Compared with a methyl group, a *t*-butyl group is much larger. We may speculate that the steric effect plays a role in the internal rotation of *t*-butyl groups. The *t*-butyl group in **TB** has a six-fold rotational potential curve with a barrier of 0.62 kcal/mol [12]. The transition state is the conformation with one of the Cm-Ct bonds perpendicular to the benzene ring. The intramolecular interaction between the methyl groups and the hydrogen atoms on the C2 and C6 of the ring is found to contribute to the barrier. The evidence is that the two non-perpendicular methyl groups change their orientation dihedrals by 5° from the ground state (staggered) conformation. The relaxation is presumably to balance the interactions between the methyl groups and the ring hydrogen atoms. These interactions stay about same in the molecules of **3-TC**, **2,6-TTN** and **1,4-TTB**, as shown by the roughly constant six-fold components of the *t*-butyl rotational PESs.

The hyperconjugation interaction comes from the electron re-distribution between the σ and σ^* orbitals of the Ct-Cm bonds of the *t*-butyl group and the π and π^* orbitals of the aromatic ring. This is a stabilization factor for the molecules. Natural bond analysis (NBO) calculations [48-55] show that the hyperconjugation favors the conformation with the Ct-Cm bonds coplanar with and syn to the shorter ring bond, which has a higher π bond order. The contribution from the hyperconjugation to the PES has the three-fold symmetry character. If we plot the three-fold components against to the bond length difference of the two ring bonds adjacent to the *t*-butyl groups, a proportional relationship is obtained. See Figure 2.10. The straight line the best fit of the data with a slope of 18.0 ±1 kcal/mol per Å. Therefore, the contribution of the hyperconjugation to the *t*-butyl group rotational PES is also proportional to the bond order difference between the two adjacent ring bonds.

For the internal rotation of methyl groups, the largest part comes from the intra-*t*-butyl interactions, *e.g.* the hyperconjugation and steric repulsion between the methyl group and the other three C-C bonds. These interactions determine the three-fold rotational potential and the ground state and the transition state. As discussed earlier, the interactions coming from the aromatic system do not alter the three-fold shape but increase the barriers. The benzene, naphthalene, and chrysene rings have similar effects on the methyl group rotation through the repulsion interactions between the ring hydrogen atoms and the methyl groups. As seen in Figure 2.7, the constrained rotational barrier only increases slightly from 4.0 kcal/mol to 4.5 kcal/mol in the range from τ(*t*-butyl) = 60 to 120°, since the methyl hydrogens and the two ring hydrogens are far away from each other. In the range from τ(*t*-butyl) = 0 to 60° or τ(*t*-butyl) = 60 to 180°, the methyl group is close to the ring hydrogen on one side. The strong repulsion between the hydrogens causes the methyl barrier to increase to about 6.5 kcal/mol for the in-plane methyl group.

2.5.3. Rotational Coupling between *t*-Butyl and Methyl Groups

The intramolecular interaction is the driving force for the coupling between the internal rotations of the methyl and *t*-butyl groups. For the *t*-butyl group rotation, two of the three methyl groups change their orientations by 5 degrees in an alternating way in order to balance the interaction between the methyl hydrogens and the ring hydrogens. The same mechanism also explains the small magnitude of the *t*-butyl wobbling during the out-of-plane methyl rotation. However, at the methyl rotational transition state of the in-plane methyl group, there is a close contact between one of the methyl hydrogens and the ring hydrogen, which is much less than the van der Waals equilibrium distance between hydrogens (2.4 Å). To relieve the steric repulsion, the *t*-butyl group rotates by 55 degrees, for which only a small barrier of the *t*-butyl rotation needs to be overcome. This leads to the conclusion that the in-plane methyl group reorients simultaneously with the *t*-butyl group.

2.6. Conclusion

The potential energy surfaces for the *t*-butyl group rotations were studied in the isolated molecules of the three representative compounds reported in this chapter. We may draw the following conclusions based on the results and discussions above.

(1) The internal rotations of the *t*-butyl groups have small barriers in the isolated molecules. At the HF/6-311+G(d,p)//HF/6-31G(d) level, the barriers are 1.14, 1.30 and 0.68 kcal/mol in 3-*t*-butylchrysene, 2,6-di-*t*-butylnaphthalene and 1,4-di-*t*-butylbenzene, respectively. The ground state is the conformation with one of the Ct-Cm bond coplanar with and syn to the shorter ring bond adjacent to the *t*-butyl group.

(2) The *t*-butyl rotational PESs can be decomposed into the three-fold and six-fold components. The six-fold components are approximately the same in the three compounds and the three-fold components are proportional to the bond length difference between the two ring C-C bonds adjacent to the *t*-butyl group.

(3) The barriers to the out-of-plane methyl groups in the three compounds are nearly the same, about 4.00 kcal/mol at the HF/6-311+G(d,p)//HF/6-31G(d) level. The barriers to the in-plane methyl groups are 5.08, 5.38 and 4.17 kcal/mol in 3-*t*-butylchrysene, 2,6-di-*t*-butylnaphthalene and 1,4-di-*t*-butylbenzene, respectively. The difference is accounted by the energy difference between the ground state and the intermediate state for the *t*-butyl rotation in the three compounds.

(4) The *t*-butyl rotation is coupled with about 5° reorientations of the methyl groups. The out-of-plane methyl rotation is also accompanied by a 5° reorientation of the *t*-butyl group.

(5) The coupling between the in-plane methyl rotation and the *t*-butyl rotation is strong. The *t*-butyl rotation is required to occur with the in-plane methyl rotation to avoid the intramolecular repulsion between the ring hydrogen and the methyl hydrogens.

2.7. References

[1] Rheingold, AL; Figueroa, JS; Dybowski, C; Beckmann, PA. Superlattices, polymorphs and solid-state NMR spin–lattice relaxation (T_1) measurements of 2,6-di-*tert*-butylnaphthalene. *Chem. Comm.*, 2000, 651-652.

[2] Beckmann, PA; Paty, C; Allocco, E; Herd, M; Kuranz, C; Rheingold, AL. The relationship between crystal structure and methyl and *t*-butyl group dynamics in van der Waals organic solids. *J. Chem. Phys.*, 2004, 120(11), 5309-5314.

[3] Beckmann, PA; Fusco, FA; O'Neill, AE. Proton spin-lattice relaxation and intramolecular reorientation in solids. *J. Magn. Reson.*, 1984, 59, 63-70.

[4] Beckmann, PA; Buser, CA; Gullifer, K; Mallory, FB; Mallory, CW; Rossi, GM; Rheingold, AL. Methyl and *t*-butyl group reorientation in planar aromatic solids: Low-frequency nuclear magnetic resonance relaxometry and X-ray diffraction. *J. Chem. Phys.*, 2003, 118(24), 11129-11138.

[5] Beckmann, PA; Burbank, KS; Clemo, KM; Slonaker, EN; Averill, K; Dybowski, C; Figueroa, JS; Glatfelter, A; Koch, S; Liable-Sands, LM; Rheingold, AL. ^{1}H nuclear magnetic resonance spin-lattice relaxation, ^{13}C magic-angle-spinning nuclear magnetic resonance spectroscopy, differential scanning calorimetry, and X-ray diffraction of two polymorphs of 2,6-di-*tert*-butylnaphthalene. *J. Chem. Phys.*, 2000, 113, 1958-1965.

[6] Sidhu, PS; Penner, GH; Jeffrey, KR; Zhao, B; Wang, ZL; Goh, I. Solid-state NMR study of guest molecule dynamics in 4-alkyl-*tert*-butylbenzene/ thiourea inclusion compounds. *J. Phys. Chem. B*, 1997, 101, 9087-9097.

[7] Penner, GH; Polson, JM; Stuart, C; Ferguson, G; Kaitner, B. Deuterium NMR and X-ray crystallographic studies of guest and host motions in the thiourea/ 1,4-di-*tert*-butylbenzene inclusion compound. *J. Phys. Chem.*, 1992, 96, 5121-5129.

[8] Yamamoto, G; Oki, M. *Tetrahedron Lett.*, 1986, 49.

[9] Yamamoto, G; Oki, M. *Bull. Chem. Soc., Jpn.*, 1986, 59, 3597.

[10] Stevenson, P., E; Bhat, G; Bushweller, CH; Anderson, WG. Experimental and theoretical studies of the rate of *tert*-butyl rotation in 2-*tert*-butyl-1,3-diheteroatomic rings. The gem-dialkyl effect, ring geometry, and lone pair orientation as factors in conformational dynamics. *J. Am. Chem. Soc.*, 1974, 96, 1067-1071.

[11] Riddelll, FG; Rogerson, M. The double *tert*-butyl rotation in solidcis-2-*tert*-butyl-5-(*tert*-butylsulphonyl)-1,3-dioxane. *Magn. Reson. Chem.*, 1997, 35, 333-336.

[12] Richardson, PR; Chapman, M., A; Wilson, DC; Bates, SP; Jones, AC. The nature of conformational preference in a number of p-alkylphenols and p-alkyl benzenes. *Phys. Chem. Chem. Phys.*, 2002, 4, 4910-4915.

[13] Oki, M. *Applications of Dynamic NMR Spectroscopy to Organic Chemistry*. VCH Publishers: Deerfield Beach, FL, 2005.

[14] Lomas, JS; Adenier, A. Rate and equilibrium effects of *tert*-alkyl groups on rotamerization in 3,4-alkylenedioxy-, 3-alkoxy- and 3,4-dialkoxy-2-thienyl(di-*tert*-alkyl)methanols: an IR, NMR, kinetics and MM study. *J. Chem. Soc., Perkin,* 2, 2002, 71264-71270.

[15] Jackman, LM; Cotton, FA. *Dynamic NMR Spectroscopy*. Academic Press: New York, 1975.

[16] Berg, U; Liljefors, T; Roussel, C; Sandstrom, J. *Acc. Chem. Res.*, 1985, 18, 80.

[17] Mate, B; Suenram, RD; Lugez, C. Microwave studies of three alkybenzenes: Ethyl, *n*-propyl, and *n*-butylbenzene. *J. Chem. Phys.*, 2000, 113(1), 192-199.

[18] Speakman, LD; Papas, B., N; Woodcock, HL; Schaefer, HFI. The microwave and infrared spectroscopy of benzaldehyde: Conflict between theory and experimental deductions. *J. Chem. Phys.*, 2004, 120 (9), 4247-4250.

[19] Seeman, JI; Paine, JBI; Secor, HV; Im, H.-S; Bernstein, ER. Supersonic jet studies of alkyl-substituted pyrazines and pyridines. Minimum energy conformations and torsional motion. *J. Am. Chem. Soc.*, 1992, 114, 5269-5280.

[20] Yanborisov, VM; Zaitesev, MY; Monakov, YB. Determination of internal rotation barrier. *Russian J. App. Chem.*, 2001, 74(12), 1942-1944.

[21] Benassi, R; Taddei, FA. Theoretical approach to the factorization of the effects governing the barrier for internal rotation around the C(sp2)-C(sp3) bond into α-substituted toluenes. *J. Mol. Struct. (Theochem)*, 1997, 418, 59-71.

[22] Stevenson, P., E; Bhat, G; Bushweller, CH; Anderson, WG. Experimental and theoretical studies of the rate of *tert*-butyl rotation in 2-*tert*-butyl-1,3-diheteroatomic rings. The gem-dialkyl effect, ring geometry, and lone pair orientation as factors in conformational dynamics. *J. Am. Chem. Soc.*, 1974, 96, 1067-1071.

[23] Richardson, PR; Chapman, M., A; Wilson, DC; Bates, SP; Jones, AC. The nature of conformational preference in a number of *p*-alkylphenols and *p*-alkyl benzenes. *Phys. Chem. Chem. Phys.*, 2002, 4, 4910-4915.

[24] Pacansky, J; Yoshimine, M. Theoretical studies on the barriers for internal rotation of the methyl groups in the *tert*-butyl radical. *J. Phys. Chem.*, 1986, 90, 1980-1983.

[25] Conn, KG; Beckmann, PA; Mallory, CW; Mallory, FB. Methyl reorientation in methylphenanthrenes. I. Solid state proton spin-lattice relaxation in the 3-methyl, 9-methyl and 3,9-dimethyl systems. *J. Chem. Phys.*, 1987, 87 (1), 20-27.

[26] Schiebel, P; Amoretti, G; Ferrero, C; Paci, B; Prager, M; Caciuffo, R. Coherent internal and centre of mass quantum rotation of methyl groups in the *p*-*tert*-butyl-calix[4]arene (2:1) *p*-xylene supramolecular complex. *J. Phys.: Condens. Matter*, 1998, 10, 2221-2231.

[27] Beckmann, PA; Fusco, FA; O'Neill, AE. Proton spin-lattice relaxation and intramolecular reorientation in solids. *J. Magn. Reson.*, 1984, 59, 63-70.

[28] Beckmann, PA; Ratcliffe, CI; Dunell, B. Proton spin-lattice relaxation and methyl group rotation. *J. Magn. Reson.*, 1978, 32, 391-402.

[29] Ito, M. in *Dynamics of Excited Molecules*, edited by Kuchitsu, K. Elsevier Science, Holland, 1994, Chap. 4, 97-120.

[30] Spangler LH; Pratt, DW. in *Jet Spectroscopy and Molecular Dynamics*, edited by Hollas, J. M.; Phillips, D., Blackie Academic and Professional, Glasgow, 1995, 369-398.

[31] Liljefors, T; Allinger, NL. *J. Comput. Chem.*, 1985, 6, 478.

[32] George, P; Bock, CW; Stezowski, JJ; Hildenbrand, T; Glusker, JP. *J. Phys. Chem.*, 1988, 92, 5656.

[33] Lu, KT; Weinhold, F; Weisshaar, JC. *J. Chem. Phys.*, 1995, 102, 6787.

[34] Nakai, H; Kawai, M. *Chem. Phys. Lett.*, 1999, 307, 272.

[35] Nakai, H; Kawai, M. *J. Chem. Phys.*, 2000, 113, 2168.

[36] Goodman, L; Gu, H; Pophristic, V. *J. Chem. Phys.*, 1999, 110, 4268.

[37] Ushiyama, H; Takatsuka, K. Time-dependent probability of quantum tunneling in terms of the quasi semiclassical method. *J. Chem. Phys.*, 2003, 120 (10), 4561.

[38] Harthcock, MA; Laane, J. Calculation of two-dimensional vibrational potential energy surfaces utilizing prediagonalized basis sets and van Vleck perturbation methods. *J. Phys. Chem.*, 1985, 89, 4231-4240.

[39] Vivier-Bunge, A; H., UV; Smeyers, YG. Two-dimensional treatment of the methyl and aldehydic torsions in gas phase propanal. *J. Chem. Phys.*, 1998, 109 (6), 2279-2286.

[40] Frisch, MJ; Trucks, GW; Schlegel, HB; Scuseria, GE; Robb, MA; Cheeseman, JR; Montgomery, Jr., JA; Vreven, T; Kudin, KN; Burant, JC; Millam, JM; Iyengar, SS; Tomasi, J; Barone, V; Mennucci, B; Cossi, M; Scalmani, G; Rega, N; Petersson, GA; Nakatsuji, H; Hada, M; Ehara, M; Toyota, K; Fukuda, R; Hasegawa, J; Ishida, M; Nakajima, T; Honda, Y; Kitao, O; Nakai, H; Klene, M; Li, X; Knox, JE; Hratchian, HP; Cross, JB; Bakken, V; Adamo, C; Jaramillo, J; Gomperts, R; Stratmann, RE; Yazyev, O; Austin, AJ; Cammi, R; Pomelli, C; Ochterski, JW; Ayala, PY; Morokuma, K; Voth, GA; Salvador, P; Dannenberg, JJ; Zakrzewski, VG; Dapprich, S; Daniels, AD; Strain, MC; Farkas, O; Malick, DK; Rabuck, AD; Raghavachari, K; Foresman, JB; Ortiz, JV; Cui, Q; Baboul, AG; Clifford, S; Cioslowski, J; Stefanov, BB; Liu, G; Liashenko, A; Piskorz, P; Komaromi, I; Martin, RL; Fox, DJ; Keith, T; Al-Laham, MA; Peng, CY; Nanayakkara, A; Challacombe, M; Gill, PMW; Johnson, B; Chen, W; Wong, MW; Gonzalez, C; and Pople, JA. *Gaussian* 03, Revision B.05, Gaussian, Inc., Pittsburgh PA, 2003.

[41] Glendening, ED; Reed, AE; Carpenter, JE and Weinhold F. *NBO* Version 3.1.

[42] The X-ray crystal structures are available free of charge from the Cambridge Crystallographic Data Centre. See http://www.ccdc.cam. ac.uk/for details.

[43] Rheingold, AL. *Private Communication*, 2001.

[44] Kravers, MA; Antipin, MYu; Struchkov, YuT. *Cryst. Struct. Commun.*, 1980, 9, 955. (BUTBNZ)

[45] See http://www.ccdc.cam.ac.uk/support/documentation/conquest/ConQuest_17-1-276. html and the references therein.

[46] Allen, FH. A systematic pairwise comparison of geometric parameters obtained by X-ray and neutron diffraction. *Acta Cryst.*, 1986, *B*42, 515.

[47] Allen, FH; Kennard, O; Watson, DG; Brammer, L; Orpen, AG; Taylor, R. *J. Chem. Soc. Perkin Trans.*, 2, 1987, S1-S19.

[48] Reed, AE; Weinhold, F. *J. Chem. Phys.*, 1983, 78, 4066.

[49] Reed, AE; Weinhold, F. *J. Chem. Phys.*, 1985, 83, 1736.

[50] Reed, AE; Weinstock, RB; Weinhold, F. *J. Chem. Phys.*, 1985, 83, 735.

[51] Reed, AE; Curtiss, LA; Weinhold, F. *Chem. Rev.*, 1988, 88, 899.

[52] Weinhold, F; Carpenter, JE. *Plenum*, 1988, 227.

[53] Carpenter, JE; Weinhold, F. *J. Mol. Struct.* (*Theochem*), 1988, 169, 41.

[54] Carpenter, JE. *PhD thesis*, University of Wisconsin, Madison, WI, 1987.

[55] Foster, JP; Weinhold, F. *J. Am. Chem. Soc.*, 1980, 102, 7211.

Chapter 3

INTERNAL ROTATIONS OF AROMATIC HYDROCARBONS IN THE CRYSTALLINE SOLID STATE

3.1. ABSTRACT

Internal rotations of the *t*-butyl groups and their constituent methyl groups in the crystalline solids of 3-*t*-butylchrysene (**3-TC**), 2,6-di-*t*-butylnaphthalene (**2,6-TTN**) and 1,4-di-*t*-butylbenzene (**1,4-TTB**) are studied in this chapter. Clusters based on the X-ray diffraction structures with computed hydrogen positions are used to simulate the local environment of the *t*-butyl groups in the crystalline solids. The barriers to the *t*-butyl rotations are found to be 6.26, 5.09 and 5.24 kcal/mol in the clusters of **3-TC**, **2,6-TTN** and **1,4-TTB**, respectively, computed at the B3LYP/6-31G(d) level. About 70% to 90% of the barriers are from intermolecular interactions with a few closest neighbors. The barriers to the methyl rotations are close to those in the isolated molecules, though intermolecular interactions are found to affect the potential energy surfaces slightly in the clusters. The internal rotations of the *t*-butyl and the constituent methyl groups are independent from each other. The synchronized rotation of the in-plane methyl and the *t*-butyl groups that was found in the isolated molecules is not energetically favorable in the clusters, though the reorienting of the methyl groups by a few degrees is allowed. There is no evidence from this work for the geared rotation of pairs of *t*-butyl groups from adjacent molecules in the clusters, but neighboring *t*-butyl groups are found to undergo temporary rotation by a few degrees from their equilibrium positions during the full rotation of the central *t*-butyl group in the clusters.

3.2. INTRODUCTION

In this chapter, internal rotation for the *t*-butyl groups in the crystalline solids is studied. Reports of computational studies of internal rotation in solids have remained scarce, though there are intensive theoretical studies of the internal rotation in the isolated molecules during the past decades [1-20]. The difficulties lie in the accuracy of computational results, especially in modeling intermolecular interactions in molecular crystals [21-22], and also in the huge burden of computational cost required in the *ab initio* simulation of molecular crystals [23-27]. However, recent development of the theory on molecular crystal simulation and innovation of computer hardware and software make it possible to model molecular crystals using quantum mechanical electronic structure calculations [27-35]. The work presented in this chapter is the first study devoted to modeling internal rotation in crystalline solids using the *ab initio* theory and the density functional theory.

There are two basic computational models for simulating solids: periodic boundary condition calculations and cluster models [27, 36-37]. The former model is necessary for computing the bulk properties of a crystal, *e.g.* the lattice energy and cohesive energy. The cluster model has been extensively used to study defects in solids and chemical reactions in the solid state. This model is more appropriate for the study of the internal rotation, since the rotational dynamics of each *t*-butyl group in the crystalline solids is most likely independent though there is concern of the coupled motions in the solids [38-42]. (Periodic boundary conditions on the other hand will enforce all the *t*-butyl groups connected by space group symmetry to rotate simultaneously, which is unrealistic energetically.) In a cluster, the crystal environment can be viewed as the host for the *t*-butyl group. However, there is an edge effect: the edge molecules of a cluster are in a different environment from that of the central ones. This is inevitable for a limited size cluster. To avoid this problem, the supercell approach is an alternative choice, in which *t*-butyl groups are enclosed in a sufficiently large supercell consisting of several unit cells, and the supercell is periodically repeated throughout space. However, this approach is impractical for this study due to limit of computational cost. In order to avoid the simultaneous rotation (which is an artificial rotational coupling) of two *t*-butyl groups in neighboring cells in the periodic boundary condition calculation, the supercell needs to be large enough. For the molecular crystals in this study, with hundreds of atoms in a unit cell, the computational cost becomes impractical under the present computational conditions. On the other hand, if a cluster is big enough, the edge effects can be ignored.

The ONIOM (acronym for 'Our owN n-layered Integrated molecular Orbital + molecular mechanics Method') method, which is a hybrid model to treat the core atoms quantum mechanically and the outer layers classically by molecular mechanics, has been popular in the study of chemical reactions in the solid state [43-49]. We prefer to use the full quantum mechanics calculation in this study. As shown in the previous chapter, the internal rotation is a process with low barriers of only a few kcal/mol. The quantum mechanics (QM) calculation gives more reliable results. Also the QM approach is well defined, and systematic improvement is possible.

In solid state chemistry modeling, it is important to evaluate the intermolecular interactions accurately. The Hartree-Fock (HF) model is limited in the treatment of the electron correlation effect, which is important in calculating the intermolecular interactions. The hybrid density functional theory, B3LYP, which captures more electron correlation effect, is adopted in this study. The model approximates the short range to intermediate range intermolecular interactions well though there is a well-known problem regarding long range interactions. Furthermore, the model is more cost-effective compared with other post-HF methods. Theoretical studies on the crystal structure of aspirin showed that the B3LYP optimized structure is in better agreement with the experimental crystal structure than the second-order Møller-Plesset perturbation theory (MP2) optimized structure [26]. Recent developments on the local electron correlation methods, *e.g.* local-MP2, local-CCSD, make it possible to apply them in large cluster models and in the periodic boundary condition calculation [28-35]. Comparison between the performance of the density functional theory and the local version of the post-HF methods is an interesting project for the future.

To build the clusters, X-ray diffraction structures of the crystals, which have been measured previously [39, 50-52], are used, though the neutron diffraction structure is more appealing since hydrogen atoms are accurately determined as well [53-54]. (Neutron diffraction structures of the crystals studied here are unavailable.) The neighboring molecules in the first interaction shell of the *t*-butyl groups are included in the clusters. Two kinds of potential energy surfaces (PESs) are then calculated in the clusters. The first kind is the rigid rotation PES, in which all the structural parameters remain fixed during the internal rotation process. The other kind is the partially relaxed rotation PES. Selective structural parameters are allowed to be optimized during the internal rotation process. The rigid rotation PES allows an easy scheme to evaluate the role of intermolecular interactions in the crystals. The partially relaxed rotation PES simulates the rotation process more accurately, and the role of the structural relaxation can be investigated by this calculation [55-56]. In addition, the possibility of *intra*molecular or *inter*molecular rotational coupling can be investigated as well.

3.3. Methods

The cluster models of 3-*t*-butylchrysene [51], 2,6-di-*t*-butylnaphthalene [39] and 1,4-di-*t*-butylbenzene [52] were built from the X-ray crystal structure of the compounds with home-written codes. The models included all the molecules with any atom falling into a sphere of radius 7.5 Å or 7.2 Å with its center located at the center carbon atoms (Ct) of the *t*-butyl groups. The radii were chosen to include a sufficient number of neighboring molecules to reasonably model the local environment of the central *t*-butyl groups. The structural parameters of hydrogen atoms in the clusters were optimized at the HF/3-21G level.

The potential energy surfaces were calculated at the HF/3-21G and B3LYP/6-31G(d) levels using the *Gaussian03* software suite [57] under the Redhat Linux OS on dual-processor Dell Precision 650 workstations. For the partially relaxed potential energy surfaces of the clusters, selective geometry parameters were optimized at the HF/3-21G level, followed by a single point calculation at the B3LYP/6-31G(d) level. The basis set superposition error (BSSE) was not corrected in all the calculations due to the following reasons: (1) The error is largely cancelled in calculating the rotational barriers; (2) large BSSE and consequent counterpoise correction would be expected to adversely affect the reliability of the potential energy surface; and (3) the counterpoise correction of the BSSE is somewhat controversial [58-65]. The potential energy surfaces were drawn with *Origin 7.0* (OriginLab Co., Northampton, MA.). The Renka-Cline gridding method was used to generate the two-dimensional potential energy surfaces. Other graphs were created with the aid of *Mathematica 5.0* (Wolfram Research, Inc., Champaign, IL.).

3.4. Results

3.4.1. Cluster Model Building

Crystal structures of the compounds, 3-*t*-butylchrysene (**3-TC**), 2,6-di-*t*-butylnaphthalene (**2,6-TTN**) and 1,4-di-*t*-butylbenzene (**1,4-TTB**) have been measured by the X-ray diffraction method and deposited in the Cambridge Crystal Database [50] with the entry codes of QEFBAE [51], KOKQUW [39] and BUTBNZ [52], respectively. There is only one unique molecule in the unit cell for each of these three crystals. The molecular structures of the compounds were presented in the previous chapter. Parameters of the unit cells are collected in Table 3.1.

To simulate the local environment for the internal rotation of the *t*-butyl groups in the crystalline solids, cluster models were built from the X-ray crystal structures of the compounds. Since there is only one unique molecular site in the three crystals, and the inversion symmetry is conserved in the crystals of **2,6-TTN** and **1,4-TTB**, all the *t*-butyl groups in each crystal share a common local environment except those on the crystal surfaces. Only the internal rotation of the *t*-butyl groups in this common environment is studied here, which is presumably dominant in the experimental observations. The dynamics on the surfaces of the clusters are more complicated and information on the surface structure is required.

In the crystals, an interaction sphere is defined. The center of the sphere is set on the central carbon atom of a *t*-butyl group (Ct). The radius is chosen to include a sufficient number of neighboring molecules to reasonably simulate the local environment of the *t*-butyl group.

All the neighbor molecules with any atom falling into the sphere are included in the cluster. The critical radius was chosen to be 7.5 Å in the crystal of **3-TC**, and 13 molecules are included in the cluster. For the crystal of **2,6-TTN**, the radius was chosen to be 7.5 Å, and the cluster includes only 11 molecules. The critical radius was set as 7.2 Å for **1,4-TTB**, and the cluster includes 13 molecules. Structures of the three clusters are shown in Figures 3.1, 3.2 and 3.3. For the convenience of description, the *t*-butyl groups are labeled with numbers. The number '1' always refers to the central *t*-butyl group, for which the internal rotation is studied.

The order of other numbers is arbitrary. The clusters are the primary model to study the internal rotation of the *t*-butyl and methyl groups in the crystals of the three compounds. Smaller clusters built from the primary clusters are also used for some purposes, and they are discussed later.

The positions of hydrogen atoms are difficult to locate by X-ray crystallography. The C-H bond lengths that are obtained by X-ray crystallography are significantly shorter than the real values [53,54,56a].

To solve this problem, the hydrogen positions in our three compounds were optimized at the HF/3-21G level in the clusters with the carbon frames frozen. (In some clusters, parallel studies were carried out using the C-H bond lengths reported from X-ray measurements. These studied showed that the results reported in this chapter are not affected significantly by this alternative choice for C-H bond lengths.) The C-H bond lengths obtained from these calculations are similar to those obtained for the isolated molecules optimized at the HF/6-31G(d) level.

Table 3.1. Crystal unit cell parameters of the compounds, measured by the X-ray diffraction method

Parameter	3-*t*-butylchrysene (3-TC)	2,6-di-*t*-butyl-naphthalene (2,6-TTN)	1,4-di-*t*-butyl benzene (1,4-TTB)
code[a]	QEFBAE	KOKQUW	BUTBNZ
space group	$P2_1/n$	$P2_1/c$	$P2_1/n$
a, Å	9.7207(12)	11.3649(4)	6.312(2)
b, Å	6.1898(7)	9.9406(3)	9.983(2)
c, Å	26.791(3)	6.6728(2)	9.706(2)
α,°	90	90	90
β,°	97.011(2)	93.943	96.02
γ,°	90	90	90
V, Å^3	1599.9(3)	752.068	608.228
Z	4	2	2
T, K[b]	173	223	153

[a] Entry code in the Cambridge Crystal Database. [b] Measurement temperature.

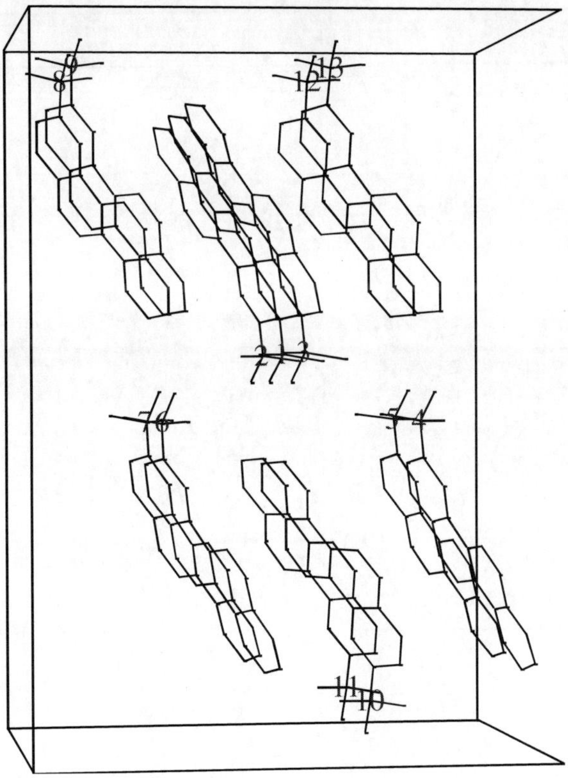

Figure 3.1. The 13-molecule cluster model of 3-*t*-butylchrysene. For clarity, hydrogen atoms are not shown. The central *t*-butyl group is labeled by '1' while the order of the labeling of the other *t*-butyl groups is arbitrary.

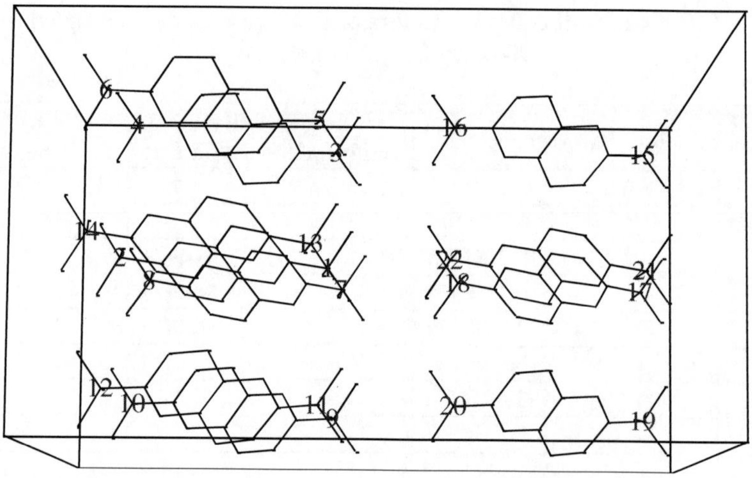

Figure 3.2. The 11-molecule cluster model of 2,6-di-*t*-butylnaphthalene. For clarity, hydrogen atoms are not shown. The central *t*-butyl group is labeled by '1' while the order of the labeling of the other *t*-butyl groups is arbitrary.

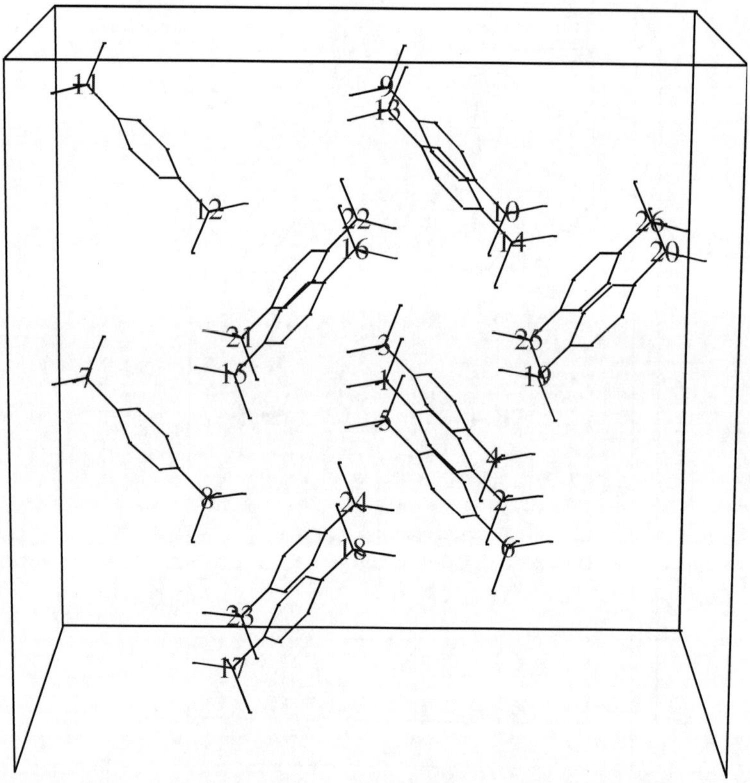

Figure 3.3. The 13-molecule cluster model of 1,4-di-*t*-butylbenzene. For clarity, hydrogen atoms are not shown. The central *t*-butyl group is labeled by '1' while the order of the labeling of the other *t*-butyl groups is arbitrary.

3.4.2. Rigid Rotation of *t*-Butyl Groups in the Clusters

The PESs for the rigid rotation of *t*-butyl and methyl groups in the clusters were first calculated. In this calculation scheme, only the internal rotation coordinates, which are the orientation dihedrals of the *t*-butyl and methyl groups as defined in the last chapter, were altered and all the other structural parameters were frozen. Two theoretical chemistry models, HF/3-21G and B3LYP/6-31G(d), were used. The former one is faster, but less accurate, and the latter one captures more electron correlation effect, which is critical to evaluate intermolecular interactions. Presumably, internal rotation of the *t*-butyl groups in the crystalline solids is strongly dependent on the short range to intermediate range of the intermolecular interactions. The B3LYP/6-31G(d) method is also faster and scales more slowly with respect to the size of the molecular system, compared with the Hartree-Fock model and post-Hartree-Fock methods with a similar basis set. This performance advantage of the B3LYP/6-31G(d) method is especially significant for large molecular systems like the clusters in this thesis. The rigid rotation PESs for the central *t*-butyl groups in the clusters are shown in Figure 3.4 calculated at the HF/3-21G level and in Figure 3.5 at the B3LYP/6-31G(d) level. (See the previous chapter for the definitions of τ(*t*-butyl) and τ(methyl).) The PESs obtained at the two levels have a similar shape but different heights. For the molecular systems composed with only carbon and hydrogen atoms as studied in this chapter, the B3LYP/6-31G(d) method, which is a hybrid density functional theory (DFT), always gives a lower barrier than the HF/3-21G method. This performance characteristic of the DFT has been familiar to computational chemists. Usually, the DFT results have a better accuracy than the Hartree-Fock model. (See ref 65a and references therein.)

In the clusters of 3-*t*-butylchrysene (**3-TC**) and 2,6-di-*t*-butylnaphthalene (**2,6-TTN**), the ground state orientations of the central *t*-butyl groups are found to have τ(*t*-butyl) nearly 0°, which agree well with the thermal equilibrium orientations found in the X-ray diffraction experiments (τ(*t*-butyl) = 1.7° in **3-TC**, and 0.3° in **2,6-TTN**). In the crystal of **1,4-TTB**, the orientation of the *t*-butyl groups is measured as τ(*t*-butyl) = 8.3° by X-ray diffraction. This same orientation is found to be the computed ground state for the central *t*-butyl group in the cluster as well. See Figures 3.4 and 3.5. (The sign of the orientation dihedral is meaningful in the crystal. The same convention is used throughout.) The rotational transition states in the clusters are reached by rotating the *t*-butyl groups around their Ct-Ca bonds by 60°. For **3-TC** and **2,6-TTN**, the transition state is the conformation with the central *t*-butyl group at τ(*t*-butyl) = 60.0°, and for **1,4-TTB**, τ(*t*-butyl) = 68.3°. In the order of **3-TC**, **2,6-TTN** and **1,4-TTB**, the barrier calculated at the B3LYP/6-31G(d) level decreases from 6.70 to 4.51 kcal/mol.

As was found in the isolated molecules, the PESs can be fit with the sum of a three-fold and a six-fold cosine potential term, $V(\tau)=\frac{V_3}{2}(1-\cos 3\tau)+\frac{V_6}{2}(1-\cos 6\tau)$. The exception is that the PES of **1,4-TTB** is fitted with the function, $V(\tau)=\frac{V_3}{2}(1-\cos 3(\tau-8.3^\circ))+\frac{V_6}{2}(1-\cos 6(\tau-8.3^\circ))$. The phase shift reflects the ground state orientation of the *t*-butyl groups in the crystal. The least-square best fit results for the three-fold and six-fold components are tabulated in Table 3.2. The best-fit curves are plotted in Figures 3.4 and 3.5 with solid lines. More discussions on the fitting results can be found in the discussion section.

Table 3.2. The least-square best fit results of the three-fold (V_3) and six fold (V_6) components of the rigid rotation potentials of *t*-butyl groups in the clusters, computed at the HF/3-21G and B3LYP/6-31G(d) levels

Compound	HF/3-21G	
	V_3, kcal/mol	V_6, kcal/mol
3-*t*-butylchrysene[a]	8.63 ± 0.15	1.05 ± 0.17
2,6-di-*t*-butylnaphthalene[a]	6.97 ± 0.07	0.87 ± 0.08
1,4-di-*t*-butylbenzene[b]	5.91 ± 0.07	0.17 ± 0.09
	B3LYP/6-31G(d)	
3-*t*-butylchrysene[a]	6.70 ± 0.02	1.00 ± 0.02
2,6-di-*t*-butylnaphthalene[a]	5.34 ± 0.05	0.94 ± 0.06
1,4-di-*t*-butylbenzene[b]	4.35 ± 0.07	0.10 ± 0.08

[a]Fitted with the function, $V(\tau) = \frac{V_3}{2}(1-\cos 3\tau) + \frac{V_6}{2}(1-\cos 6\tau)$.

[b]Fitted with the function, $V(\tau) = \frac{V_3}{2}(1-\cos 3(\tau - 8.3^\circ)) + \frac{V_6}{2}(1-\cos 6(\tau - 8.3^\circ))$.

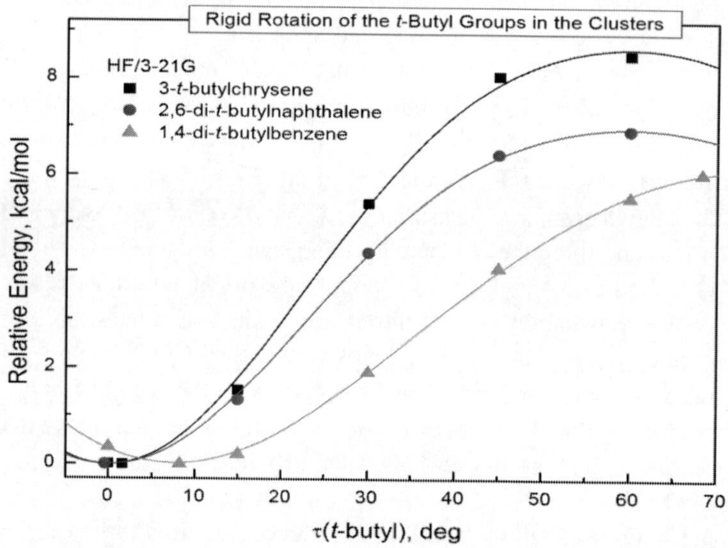

Figure 3.4. The rigid rotational potential energy surfaces for the internal rotation of *t*-butyl groups in the clusters, calculated at the HF/3-21G level. The solid lines are the least-square best fit with the function, $V(\tau) = \frac{V_3}{2}(1-\cos 3\tau) + \frac{V_6}{2}(1-\cos 6\tau)$ for 3-TC, 2,6-TTN, and with the function, $V(\tau) = \frac{V_3}{2}(1-\cos 3(\tau - 8.3^\circ)) + \frac{V_6}{2}(1-\cos 6(\tau - 8.3^\circ))$ for 1,4-TTB. The best-fit V_3 and V_6 components are listed in Table 3.2.

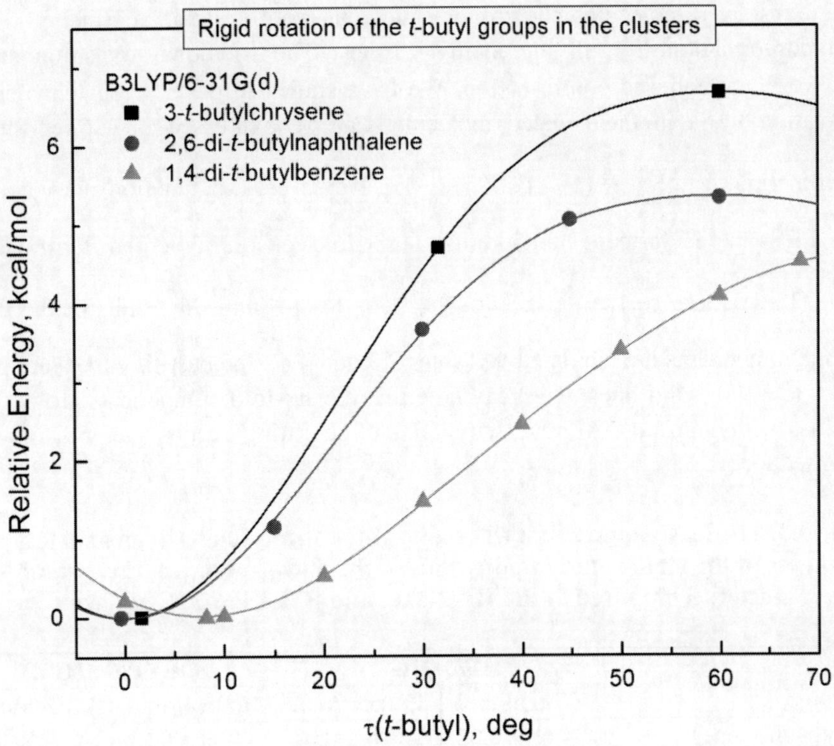

Figure 3.5. The rigid rotational potential energy surfaces for the internal rotation of *t*-butyl groups in the clusters, calculated at the B3LYP/6-31G(d) level. The solid lines are the least-square best fit with the function, $V(\tau)=\frac{V_3}{2}(1-\cos 3\tau)+\frac{V_6}{2}(1-\cos 6\tau)$ for 3-TC and 2,6-TTN, and with the function, $V(\tau)=\frac{V_3}{2}(1-\cos 3(\tau-8.3^\circ))+\frac{V_6}{2}(1-\cos 6(\tau-8.3^\circ))$ for 1,4-TTB. The best-fit V_3 and V_6 components are listed in Table 3.2.

3.4.3. Rigid Rotation of the Constituent Methyl Rotations in the Clusters

The potential energy surfaces for the rigid rotation of the constituent methyl groups on the central *t*-butyl groups were also calculated at the two levels. The PESs for the out-of-plane methyl rotations are shown in Figures 3.6 and 3.8. Because the ground state orientation of the central *t*-butyl group in **1,4-TTB** is not 0°, the two out-of-plane methyl groups are not symmetric about the benzene ring plane. The results for the rigid rotation of both the groups are shown in Figure 3.6.

The rigid rotation PESs for the in-plane methyl groups in **3-TC** and **2,6-TTN** and the nearly in-plane methyl group in **1,4-TTB** (labeled by τ(*t*-butyl) = 8.3°) are shown in Figures 3.7 and 3.9. The higher level calculation, B3LYP/6-31G(d), was only done in the clusters of **3-TC** and **2,6-TTN**. The conclusion found from the two compounds can be applied to **1,4-TTB**.

As in the isolated molecules, the ground state is the conformation at τ(methyl) = 60°, the staggered conformation, for all the methyl groups. The rotational transition state is at τ(methyl) = 0°, the eclipsed conformation. We have shown in the previous chapter that the methyl rotation PESs in the isolated molecules can be expressed in a three-fold cosine potential function, $V(\tau)=\frac{V_3}{2}(1-\cos 3(\tau-\frac{\pi}{3}))$. However, the function does not fit well with the PESs here. Instead, an additional six-fold cosine potential term is needed, $V(\tau)=\frac{V_3}{2}(1-\cos 3(\tau-\frac{\pi}{3}))+\frac{V_6}{2}(1-\cos 6\tau)$. The least-square fit results of the three-fold and six-fold components are tabulated in Tables 3.3 and 3.4. The best-fit curves are plotted in Figures 3.6 to 3.9 in solid lines. One may note that the six-fold components are negative for most of the methyl groups. More discussion on the fitting results can be found in the discussion section.

Table 3.3. The least-square best fit results of the three-fold (V_3) and six fold (V_6) components of the rigid rotation potentials of the out-of-plane methyl groups in the clusters, computed at the HF/3-21G and B3LYP/6-31G(d) levels

Compound	HF/3-21G[a]		B3LYP/6-31G(d)[a]	
	V_3, kcal/mol	V_6, kcal/mol	V_3, kcal/mol	V_6, kcal/mol
3-*t*-butylchrysene	4.68 ± 0.02	-0.06 ± 0.02	4.33 ± 0.01	-0.22 ± 0.01
2,6-di-*t*-butylnaphthalene	4.94 ± 0.04	-0.47 ± 0.05	4.02 ± 0.04	-0.39 ± 0.04
1,4-di-*t*-butylbenzene[b]	5.15 ± 0.05	-0.55 ± 0.06		
1,4-di-*t*-butylbenzene[c]	5.22 ± 0.01	0.07 ± 0.01		

[a] Fitted with the function, $V(\tau)=\frac{V_3}{2}(1-\cos 3(\tau-\frac{\pi}{3}))+\frac{V_6}{2}(1-\cos 6\tau)$.

[b] The out-of-plane methyl labeled by τ(*t*-butyl) = 128.4°.

[c] The out-of-plane methyl labeled by τ(*t*-butyl) = -111.9°.

Table 3.4. The least-square best fit results of the three-fold (V_3) and six fold (V_6) components of the rigid rotation potentials of the in-plane methyl groups in the clusters, computed at the HF/3-21G and B3LYP/6-31G(d) levels

Compound	HF/3-21G[a]		B3LYP/6-31G(d)[a]	
	V_3, kcal/mol	V_6, kcal/mol	V_3, kcal/mol	V_6, kcal/mol
3-*t*-butylchrysene	6.28 ± 0.01	-0.20 ± 0.01	5.27 ± 0.03	-0.25 ± 0.03
2,6-di-*t*-butylnaphthalene	7.36 ± 0.04	-0.52 ± 0.05	5.96 ± 0.001	-0.42 ± 0.002
1,4-di-*t*-butylbenzene	6.00 ± 0.04	-0.60 ± 0.05		

[a] Fitted with the function, $V(\tau)=\frac{V_3}{2}(1-\cos 3(\tau-\frac{\pi}{3}))+\frac{V_6}{2}(1-\cos 6\tau)$.

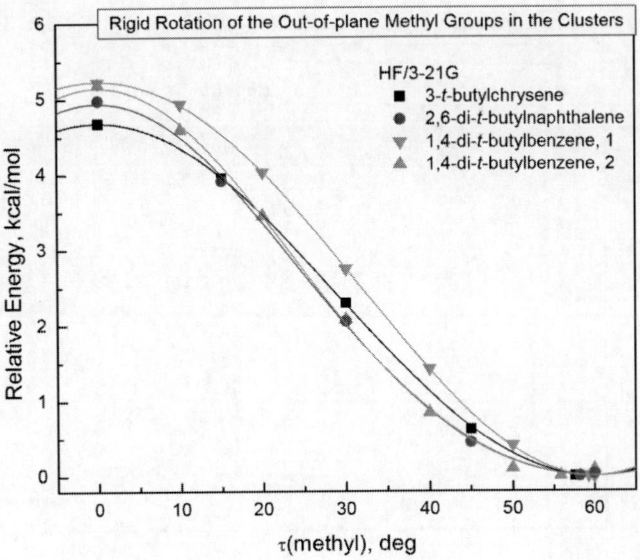

Figure 3.6. The rigid rotational potential energy surfaces for the out-of-plane methyl groups in the clusters, calculated at the HF/3-21G level. The solid lines are the least-square best fit with the function, $V(\tau)=\frac{V_3}{2}(1-\cos 3(\tau-\frac{\pi}{3}))+\frac{V_6}{2}(1-\cos 6\tau)$. The best-fit V_3 and V_6 components are listed in Table 3.3.

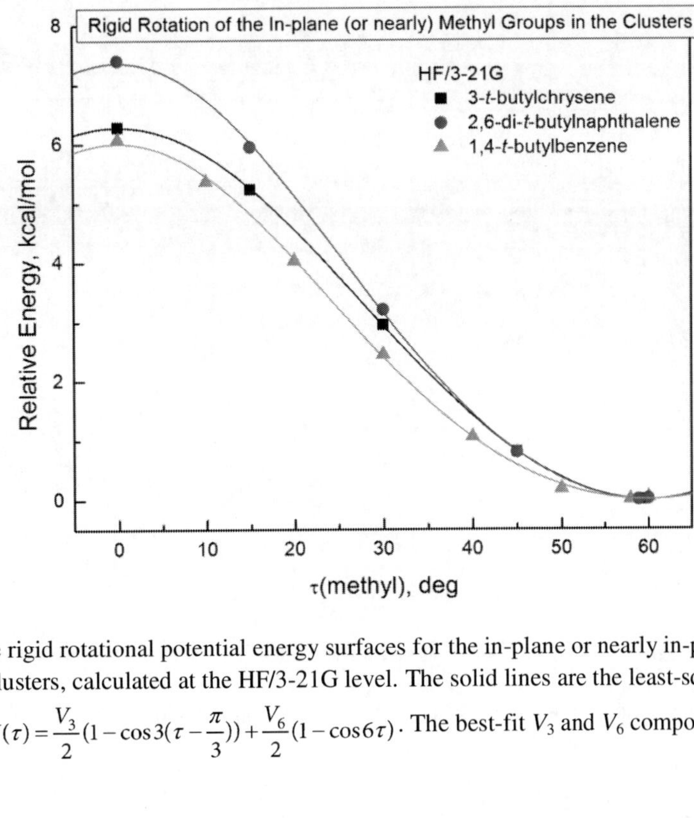

Figure 3.7. The rigid rotational potential energy surfaces for the in-plane or nearly in-plane methyl groups in the clusters, calculated at the HF/3-21G level. The solid lines are the least-square best fit with the function, $V(\tau)=\frac{V_3}{2}(1-\cos 3(\tau-\frac{\pi}{3}))+\frac{V_6}{2}(1-\cos 6\tau)$. The best-fit V_3 and V_6 components are listed in Table 3.4.

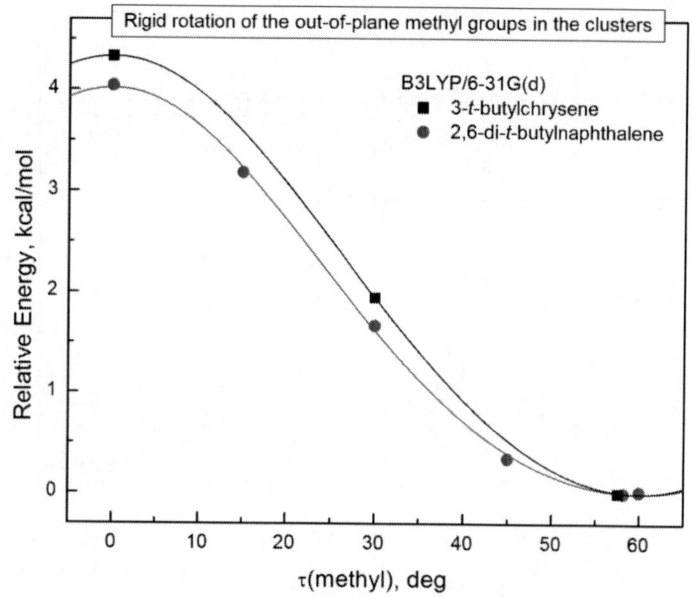

Figure 3.8. The rigid rotational potential energy surfaces for the out-of-plane methyl groups in the cluster, calculated at the B3LYP/6-31G(d) level. The solid line is the least-square best fit with the function, $V(\tau)=\frac{V_3}{2}(1-\cos 3(\tau-\frac{\pi}{3}))+\frac{V_6}{2}(1-\cos 6\tau)$. The best-fit V_3 and V_6 components are listed in Table 3.3.

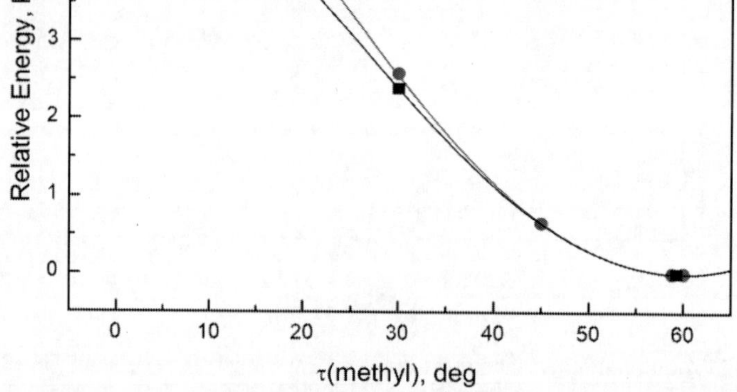

Figure 3.9 The rigid rotational potential energy surfaces for the in-plane methyl groups in the clusters, calculated at the B3LYP/6-31G(d) level. The solid lines are the least-square best fit with the function, $V(\tau)=\frac{V_3}{2}(1-\cos 3(\tau-\frac{\pi}{3}))+\frac{V_6}{2}(1-\cos 6\tau)$. The best-fit V_3 and V_6 components are listed in Table 3.4.

3.4.4. Partially Relaxed Rotation of the *t*-Butyl Groups in the Clusters

It has been found that structural relaxation plays an important role in internal rotation in the isolated molecules [55]. In the crystalline solids, besides the possible role of the intramolecular relaxation effect, it has been long argued that the geared rotation may also exist between the close-by *t*-butyl groups from the neighboring molecules. This geared rotation is one of the most interesting features of the possible intermolecular structural relaxation.

To study how structural relaxation affects the rotational processes in the crystals, we carried out the calculations of the partially relaxed potential energy surfaces in the clusters. A fully relaxed potential energy surface is impractical to obtain in the clusters due to two main reasons. The first reason is that the cluster models do not have the periodical boundary conditions that exist in a real crystal. A full relaxation will lead the clusters to fall apart and to attain the computed equilibrium distances and orientations between the molecules, which are ideal for the clusters, and presumably far away from the values in the crystals. The second reason is that the big size of the clusters makes the full optimization impractical at any levels of quantum mechanics electronic structure calculations. The clusters are composed of 11 to 13 molecules, about 300 to 500 atoms. Not only is the computational cost huge, and but also it is hard to ensure that the global minimum is achieved.

The partial relaxation scheme assumes that the carbon skeletons of the aromatic ring systems are rigid. This is reasonable since their large size and the strong packing interactions among them are expected to force them to stay at their crystal locations except for slight thermal vibrations around the equilibrium positions. The calculations for the isolated molecule of **TB** in the previous chapter also show that whether to allow the aromatic rings to relax has no big effect on the barrier and the PES.

Keeping the aromatic skeletons fixed, we allow the *t*-butyl groups and the ring hydrogen atoms in the clusters to be optimized at the HF/3-21G level. Three consecutive optimization steps are used to identify which motions are important in the relaxation. In the 1st step, only the orientation dihedrals of all the *t*-butyl groups (three τ(*t*-butyl) for each group) are allowed to optimize. (This step practically relieves the constraint on both the orientation of the *t*-butyl groups and the bond angles between the Ct-Cm bonds.) In the 2nd step, besides the *t*-butyl orientations, the dihedrals of all the methyl groups are also optimized. (This step further relieves the constraint on both the orientation of the constituent methyl groups and the bond angles between the Cm-H bonds.) In the 3rd step, not only the dihedrals but also the bond lengths and bond angles of all the *t*-butyl and methyl groups and the ring hydrogens are optimized. (Full optimization is applied to the *t*-butyl groups and ring hydrogen atoms in this step.) Single point energy calculations at the B3LYP/6-31G(d) level are then followed in the optimized clusters to obtain more accurate barriers.

In Table 3.5, the results of the partially relaxed barriers are shown. (A full three-step calculation is only done in the cluster of **3-TC** as a representative example due to the time cost concern.) In the cluster of **3-TC**, the biggest barrier decrease is found to occur in the 1st step, from 8.55 to 8.10 kcal/mol at the HF/3-21G level, and from 6.61 to 6.26 kcal/mol at the B3LYP/6-31G(d) level. During the optimization steps two and three, the barriers stay almost the same. Searching for the structural changes, we find that significant structural changes only happen in the 1st step. Two of the neighbor *t*-butyl groups (*t*-butyl groups 5 and 7 in Figure

3.1) are found to adjust their orientations by 7 to 8° as the central *t*-butyl group reorients from the ground state to the transition state.

The 1st optimization step also lowers the rotational barrier of the central *t*-butyl group from 5.32 kcal/mol to 5.09 kcal/mol in the cluster of **2,6-TTN** at the B3LYP/6-31G(d) level. No big structural changes are found except the small magnitudes of the dihedral changes of a few closest neighbor molecules as found similarly in the cluster of **3-TC**. The partial optimization treatment in the cluster of **1,4-TTB** has followed a different procedure, and it is discussed in the later section of 3.4.7.

Table 3.5. Rigid and partially relaxed rotational barriers of the central *t*-butyl groups in the clusters of the three compounds

Steps[a]	Barrier to the *t*-butyl rotation, kcal/mol					
	3-*t*-butylchrysene		2,6-di-*t*-butylnaphthalene		1,4-di-*t*-butylbenzene[d]	
	HF[b]	B3LYP[c]	HF[b]	B3LYP[c]	HF[b]	B3LYP[c]
Rigid	8.55	6.61	6.93	5.32	5.64	4.51
1st	8.10	6.26	6.68	5.09	5.24	
2nd	8.02		6.33	4.90		
3rd	8.13		4.82	3.66		

[a] See the text in Section 3.4.4 for details.
[b] HF/3-21G.
[c] B3LYP/6-31G(d).
[d] Different optimization step was adopted. See the text in Section 3.4.7 for details.

Table 3.6. Rigid and partially relaxed rotational barriers of the in-plane or nearly in-plane methyl groups on the central *t*-butyl groups in the clusters

Steps	Barrier to the in-plane methyl rotation, kcal/mol					
	3-*t*-butylchrysene		2,6-di-*t*-butylnaphthalene		1,4-di-*t*-butylbenzene	
	HF[a]	B3LYP[b]	HF[a]	B3LYP[b]	HF[a]	B3LYP[b]
Rigid	6.28	5.27	7.41	5.99	6.07	
Relaxed			6.98	5.63		

[a] HF/3-21G.
[b] B3LYP/6-31G(d).

Table 3.7. Rigid and partially relaxed rotational barriers of the out-of-plane methyl groups on the central *t*-butyl groups in the clusters

Steps	Barrier to the out-of-plane methyl rotation, kcal/mol					
	3-*t*-butylchrysene		2,6-di-*t*-butylnaphthalene		1,4-di-*t*-butylbenzene	
	HF[a]	B3LYP[b]	HF[a]	B3LYP[b]	HF[a]	B3LYP[b]
Rigid	4.68	4.33	4.98	4.04	5.22(5.19)[c]	
Relaxed			4.46	3.82		

[a] HF/3-21G.
[b] B3LYP/6-31G(d).
[c] The value in parenthesis is for the out-of-plane methyl group labeled by τ(*t*-butyl) = 128.4°.

3.4.5. Partially Relaxed Rotation of the Constituent Methyl Groups in the Clusters

The partially relaxed rotations were also calculated for the constituent methyl groups on the central *t*-butyl group in the cluster of **2,6-TTN** as a representative example. The aromatic backbones of the molecules in the cluster were fixed, but the *t*-butyl groups were optimized at the HF/3-21G level. The results are shown in Tables 3.6 and 3.7 along with the rigid rotational barriers. The partially relaxed rotation barriers are slightly lower than the rigid rotation ones due to the small magnitudes of the changes on the bond lengths, angles and dihedrals.

3.4.6. Two-dimensional Potential Energy Surface with respect to the Partially Relaxed Rotations of the *t*-Butyl and Methyl Groups in the Cluster of 3-TC

In the isolated molecules, the in-plane methyl rotation is found to occur simultaneously with the *t*-butyl group. For the *t*-butyl and out-of-plane methyl groups, weakly coupled orientation relaxation is observed. To investigate the coupling between the rotations of the *t*-butyl and the constituent methyl groups in the crystalline solid state, a two-dimensional potential energy surface with respect to the partially relaxed rotations of the *t*-butyl and the related methyl groups is calculated in a reduced-size cluster of **3-TC** as a representative example.

If we define the 3-*t*-butyl group as the head of **3-TC** and the far-end ring as the tail, there are two kinds of neighbors in the crystal. One is head-to-head, and the other is tail-to-head. In the 13-molecule cluster, there are 6 head-to-head and 6 tail-to-head neighbors for the central molecule. To reduce the computational cost, the 6 tail-to-head neighbors are removed from the cluster to calculate the two-dimensional PES. As shown later, the feasibility of using this reduced 7-molecule cluster is rationalized by the fact that the 6 tail-to-head neighbors contribute little to the *t*-butyl rotational barrier on the central molecule. With this cluster model, the partially relaxed two-dimensional PES was calculated, in which the carbon frames of the rings were fixed and the structural parameters of the *t*-butyl groups and the ring hydrogen atoms were optimized at the HF/3-21G level, followed by a single point energy calculation at the B3LYP/6-31G(d) level. The PES is shown in Figure 3.10.

The two-dimensional PES obtained in the 7-molecule cluster shows different features, compared with those calculated in the isolated molecules. In the isolated molecules of **3-TC**, the conformation at τ(*t*-butyl) = 60° and τ(methyl) = 0° is a saddle point, and the conformation at τ(*t*-butyl) = 0° and τ(methyl) = 0° is a local maximum point. The in-plane methyl rotation barrier is very high (5.08 kcal/mol) and the *t*-butyl rotation barrier is very low (1.14 kcal/mol). (See Table 2.5 of Chapter 2.) In the cluster, the barrier to the rotation of the *t*-butyl groups rises to higher than the methyl rotation barrier. For this reason, the conformation at τ(*t*-butyl) = 60° and τ(methyl) = 0° is closely a local maximum point, and the conformation at τ(*t*-butyl) = 0° and τ(methyl) = 0° becomes a saddle point. Additionally, there are two troughs parallel to the τ(methyl) axis. One is for the in-plane methyl group (τ(*t*-butyl) = 0°), and the other for the out-of-plane methyl group (τ(*t*-butyl) = 120°). In the isolated molecules, there is a short-cut minimum energy path for the in-plane methyl rotation through the saddle point at τ(*t*-butyl) = 60° and τ(methyl) = 0° (the out-of-plane methyl at the *t*-butyl rotational intermediate state). The advantage of this path is blocked in the cluster due to the high barrier to the *t*-butyl rotation.

As in the isolated molecules, the PES can be regarded as the dependence of the constrained methyl rotation on the *t*-butyl orientation (τ(*t*-butyl)).(‡) The barrier to the constrained methyl rotation, which follows a straight path along the τ(methyl) direction, is plotted against τ(*t*-butyl) in Figure 3.11. The curve has a similar shape as that in the isolated molecules. The barrier to the constrained in-plane methyl rotation is about 6.00 kcal/mol, which is slightly smaller than the value found in the isolated molecules, 6.70 kcal/mol at the HF/6-311+G(d,p)//HF/6-31G(d) level. The barrier to the constrained out-of-plane methyl rotation is about 4.00 kcal/mol in the cluster, similar as in the isolated molecules. The dependence curve was fitted with the function, $V_m(\tau_t) = V_{m0}1 + V_{m2}\cos 2\tau_t + V_{m4}\cos 4\tau_t$. The best-fit coefficients are 9.00 ± 0.18, 1.38 ± 0.26 and 1.13 ± 0.26 kcal/mol for V_{m0}, V_{m2} and V_{m4}, respectively.

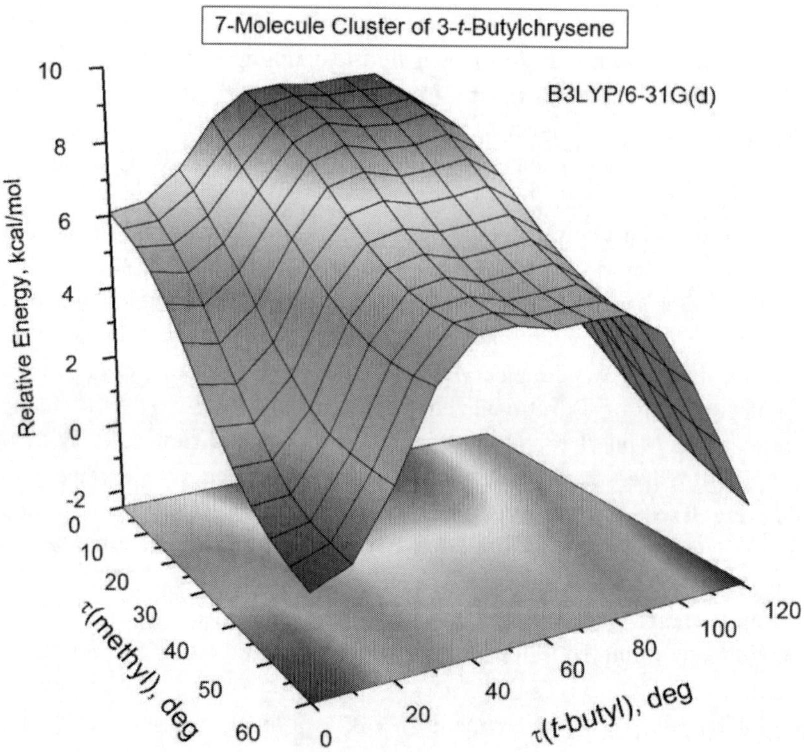

Figure 3.10. The partially relaxed potential energy surface with respect to the internal rotation coordinates of the central *t*-butyl group and the related methyl group in the 7-molecule cluster of 3-*t*-butylchrysene, calculated at the B3LYP/6-31G(d)//HF/3-21G level.

‡ An alternative view of this dependence is to fit the two-dimensional PES with the methyl rotation PES, the *t*-butyl rotation PES and the coupling term between the two, as given in the previous chapter. The fitting separates the variables, τ(*t*-butyl) and τ(methyl), and thus simplifies the problem.

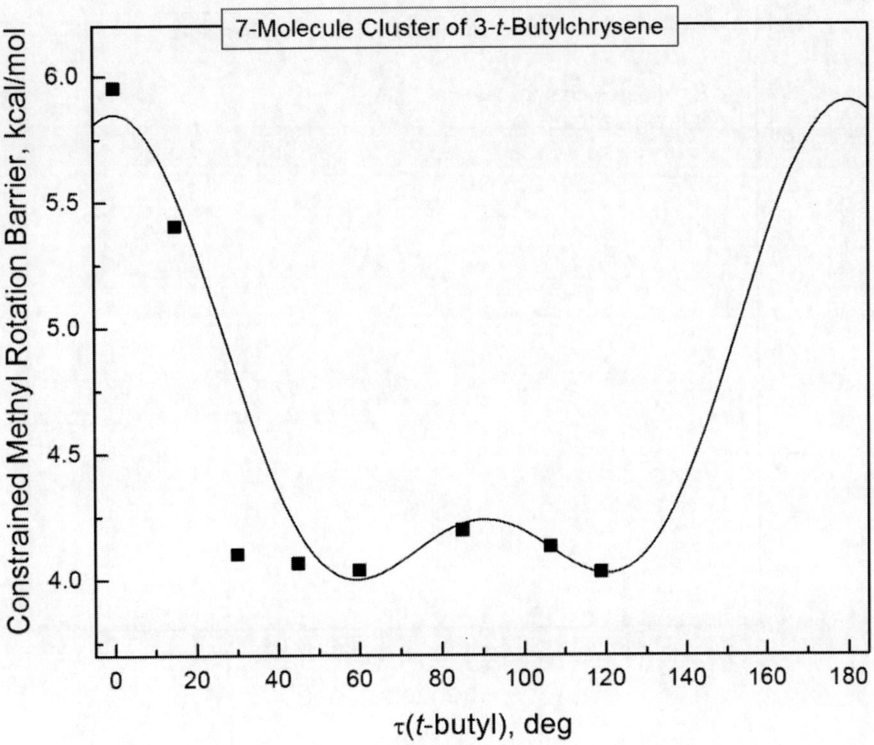

Figure 3.11. The dependence of the constrained rotation barrier of the methyl group on its orientation labeled by τ(*t*-butyl) in the 7-molecule cluster of 3-TC. The barriers are from the two-dimensional potential energy surfaces in Figure 3.10. The solid line is the least-square best fit with the function, $V_m(\tau_t) = V_{m0} 1 + V_{m2} \cos 2\tau_t + V_{m4} \cos 4\tau_t$.

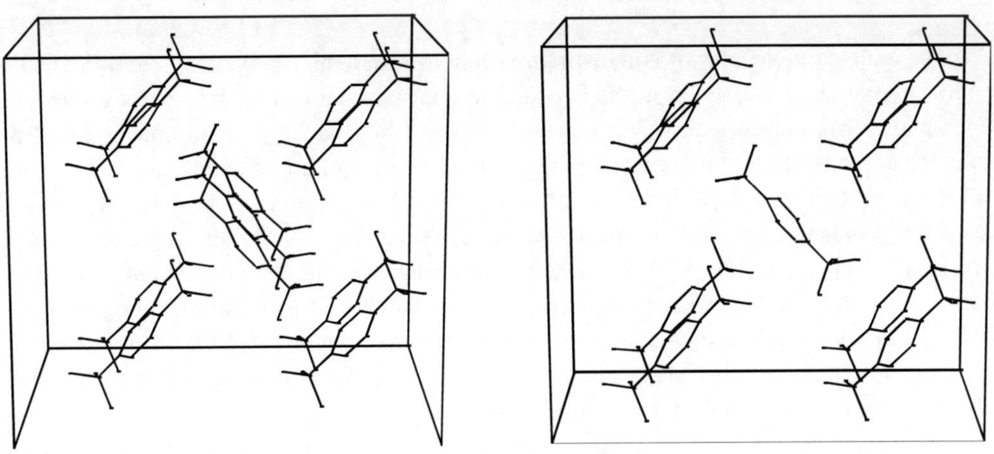

Figure 3.12. The 11-molecule cluster (upper) and 9-molecule cluster (lower) of 1,4-di-*t*-butylbenzene. For clarity, hydrogen atoms are not shown. The clusters provide a more symmetrical environment for the central molecule compared with the cluster in Figure 3.3.

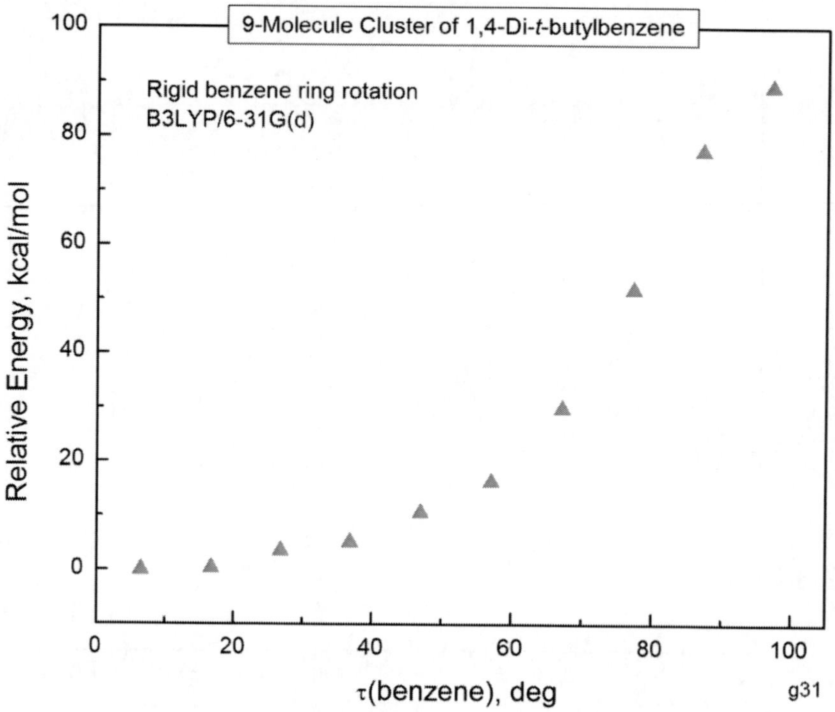

Figure 3.13. The rigid rotational potential energy surface for the benzene ring on the central molecule in the 9-molecule cluster of 1,4-TTB, calculated at the B3LYP/6-31G(d) level.

3.4.7. Rigid and Partially Relaxed Rotations of the Benzene Ring in the Cluster of 1,4-Di-*t*-butylbenzene

In the isolated molecules, it is indistinguishable to rotate the *t*-butyl group and the rest of the aromatic systems. However, in the crystal, the internal rotations of the *t*-butyl groups and the aromatic rings are not equivalent because of the crystal packing environment. In all the above calculations, the aromatic rings have been assumed to be frozen in the clusters. This assumption probably holds well for the crystals of **3-TC** and **2,6-TTN** because of the large size of the aromatic rings and the strong packing interactions. However, a benzene ring is much smaller and it may be possible for a benzene ring to rotate to some extent in a crystal environment. The NMR relaxation experiment by Penner et al. shows that the benzene ring of **1,4-TTB** can *not* rotate in the crystal of the inclusion compound of **1,4-TTB** and thiourea [66-67]. However, the possibility still remains for the benzene ring to make a small magnitude wobbling around its equilibrium position. To investigate this dynamics problem of the benzene ring, the potential energy surface was also calculated for the benzene ring rotation in a cluster model of **1,4-TTB**. Instead of using the 13-molecule cluster model, which mimics the local environment for the *t*-butyl group rotation, a 9-molecule cluster was built to represent a more complete local environment for the benzene ring in the crystal. The structure of this cluster is shown in Figure 3.12. The potential energy surface was calculated for the rigid rotation of the central benzene ring about its C1-C4 virtual axis at the B3LYP/6-31G(d) level. The PES is shown in Figure 3.13 in the range from τ(benzene) = 0 to 100°. The rotational coordinate, τ(benzene), has been defined as the dihedral of C2C1CtCm on the central *t*-butyl group. The thermal equilibrium orientation in the crystal, τ(benzene) = 8.3°, is also found as the computed ground state conformation. The transition state is the conformation with the benzene ring rotated by 90° from the ground state. The barrier to the benzene rotation is extremely high, about 89.0 kcal/mol at the B3LYP/6-31G(d) level. (In the isolated molecules, the barrier for this motion is only twice the *t*-butyl rotation barrier, *i.e.*1.36 kcal/mol.) Therefore, we may conclude it is impossible for the benzene ring to make a complete rotation in the crystal at the room temperature. However, we notice in Figure 3.13 there is a nearly flat energy range for small angles of rotation close to the ground state. For example, rotating the benzene ring by 10° from the ground state causes the barrier to increase by 0.39 kcal/mol only. A rotation of 20° increases the energy by 3.52 kcal/mol. As a result, it is likely for the benzene ring to wobble around the equilibrium orientation.

Since the benzene ring is able to wobble around the equilibrium position, does the *t*-butyl rotation couple with the benzene ring wobbling? Or is the *t*-butyl rotational barrier relieved through the wobbling motion? To answer this question, a partially relaxed rotation of the *t*-butyl group was studied on the central molecule in the 9-molecule cluster. The 8 neighbor molecules in the cluster were fixed, but the structure of the central molecule was allowed to be optimized at the HF/3-21G level, when rotating one of the *t*-butyl groups on the central molecule. (This *t*-butyl is called the 1-*t*-butyl group, and the other one is then the 4-*t*-butyl group.) For this partial relaxation calculation scheme, it is found the computed ground state is nearly the same as found in the crystal. The transition state is reached when the *t*-butyl group is rotated by 55° around its C1-Ct bond. The computed barrier decreases from 5.64 kcal/mol for the rigid rotation to 5.24 kcal/mol, calculated at the HF/3-21G level. There are three major structural changes accompanying the *t*-butyl rotation. The first major change is the rotation of the benzene ring. At the transition state, the benzene ring on the central molecule is found to rotate by 10° from the ground state orientation. The second change is the shift of the overall central molecule. It translates by 0.15 Å to the 4-*t*-butyl direction. The third change is that the 4-*t*-butyl group is driven to rotate by 6° from its original relative orientation to the benzene ring. The collective result of the above relaxations brings down the *t*-butyl rotation barrier slightly.

3.4.8. Contribution of Intermolecular Interactions to the Barriers

As shown above, the *t*-butyl rotation barriers in the clusters are much higher than those in the isolated molecules. Intermolecular interactions must contribute a significant portion of the barriers in the crystals. It is of interest to know the contribution from each neighbor. This problem can be solved easily in the cluster models. From the largest cluster of the compounds, which represents a complete local environment for the *t*-butyl rotation, we selectively remove neighbors to create a series of reduced-size clusters. The rigid rotation PES is then re-calculated in these smaller clusters. The contribution of the intermolecular interactions can be estimated by comparing the rigid rotation barrier in the complete cluster and the reduced-size cluster.

Table 3.8. Contribution of intermolecular interactions to the *t*-butyl rotation barrier in the crystal of 3-*t*-butylchrysene. In the reduced-size clusters, the rigid rotation barrier of the central *t*-butyl group was computed at the HF/3-21G and B3LYP/6-31G(d) levels. The decrease of the barrier in the smaller clusters reflects the contribution from the neighbor removed

Cluster	Rigid rotation barrier, kcal/mol		Description[a]
	HF/3-21G	B3LYP/6-31G(d)	
1	8.50	6.66	13-molecule cluster
2	8.55	6.61	Neighbors 8-13 removed
3	6.94	5.45	Neighbor 7 removed
4	5.12	4.00	Neighbors 2 and 3 removed
5	6.00	4.88	Neighbors 5 and 7 removed
6	8.58	6.64	Neighbors 4 and 6 removed
7	2.74	2.32	All neighbors removed

[a] The structure of the 13-molecule cluster is shown in Figure 3.1 and the neighbors molecules are labeled by the index of their *t*-butyl groups as in Figure 3.1.

Table 3.9. Contribution of intermolecular interactions to the *t*-butyl rotation barrier in the crystal of 2,6-di-*t*-butylnaphthalene. In the reduced-size clusters, the rigid rotation barrier of the central *t*-butyl group was computed at the HF/3-21G and B3LYP/6-31G(d) levels. The decrease of the barrier in the smaller clusters reflects the contribution from the neighbor removed

Cluster	Rigid rotation barrier, kcal/mol		Description[a]
	HF/3-21G	B3LYP/6-31G(d)	
1	6.93	5.32	11-molecule cluster
2	7.06	5.15	Neighbor 22 removed
3	6.88	4.68	Neighbors 22 and 18 removed
4	3.14	2.62	Neighbors 5 and 11 removed
5	3.53	2.81	Neighbors 3, 5, 9 and 11 removed
6	2.78	2.26	All neighbors removed

[a] The structure of the 11-molecule cluster is shown in Figure 3.2 and the neighbors molecules are labeled by the index of one of their *t*-butyl groups as in Figure 3.2. Neighbor # refers to the whole molecule on which the *t*-butyl group # is.

In Table 3.8, the rigid rotation barriers in the different size clusters of **3-TC** are shown. Since the intermolecular interaction is better evaluated at the B3LYP/6-31G(d) level, we focus on the results obtained at this level. Cluster 1 is the complete 13-molecule cluster, and the barrier is the largest, 6.66 kcal/mol. In cluster 2, the six tail-to-head neighbors are removed from cluster 1. The barrier decreases only slightly to 6.61 kcal/mol. The intermolecular interactions from the six neighbors have almost no effect to the central *t*-butyl rotation. This leads us to use this 7-molecule cluster for calculating the two-dimensional PES in the previous section, without losing any important factors in the crystal. In clusters 3 and 5, the nearest and next nearest neighbor molecules, in terms of the distance between the Ct atoms (inter-Ct distance), are removed sequentially. The barrier decreases to 5.45 kcal/mol in cluster 3 and 4.88 kcal/mol in cluster 5. From these results, it can be estimated that the intermolecular interactions from these two neighbors contribute about 27% of the overall barrier height, with the nearest one contributing about 18% and the next nearest contributing about 9%. In cluster 4, the two parallel neighbors, with a same inter-Ct distance to the central *t*-butyl group, are removed from cluster 1. The barrier is lowered to 4.00 kcal/mol. The decrease of the barrier is 2.66 kcal/mol, corresponding about 40% of the total barrier. Since the two neighbors are equivalent, each should contribute about 20%. The other two head-to-head neighbors are removed in cluster 6. The two neighbors are far away from the central *t*-butyl group, and we expect there are only weak interactions from them. As expected, there is no significant decrease in the barrier in cluster 6. Finally, in cluster 7, all the 12 neighbors are removed with only the central molecule left. The barrier is 2.32 kcal/mol[§], which is consistent with the value of 2.17 kcal/mol, obtained by subtracting the contributions of the neighbors from the overall barrier in cluster 1.

This analysis of the contribution of the intermolecular interactions can be extended to the clusters of **2,6-TTN** and **1,4-TTB**. The rigid rotational barriers in the reduced clusters have been shown in Tables 3.9 and 3.10.

Table 3.10. Contribution of intermolecular interactions to the *t*-butyl rotation barrier in the crystal of 1,4-di-*t*-butylbenzene. In the reduced-size clusters, the rigid rotation barrier of the central *t*-butyl group was computed at the HF/3-21G and B3LYP/6-31G(d) levels. The decrease of the barrier in the smaller clusters reflects the contribution from the neighbor removed

Cluster	Rigid rotation barrier, kcal/mol		Description
	HF/3-21G	B3LYP/6-31G(d)	
1	6.02	4.38	13-molecule cluster, A[a]
2	6.03	4.39	Neighbor 12 removed from A
3	6.04	4.49	Neighbors 12 and 14 removed from A
4	5.62	4.51	11-molecule cluster, B[b]
5	5.46	3.78	9-molecule cluster , C[c]
6	4.58	3.42	7-molecule model, 4 perpendicular neighbors removed from B
7	4.16		*t*-Butyl groups 15 and 16 removed from B
8	3.56		Benzene ring on neighbor 15 (or 16) removed from B
9	2.08	1.46	Neighbors 15 and 21 removed from B
10	3.89	2.82	Neighbors 15 removed from B

[a] The structure of the 13-molecule cluster (A) is shown in Figure 3.3 and the neighbors molecules are labeled by the index of one of their *t*-butyl groups as in Figure 3.3. Neighbor # refers to the whole molecule on which the *t*-butyl group # is.

[b] The structure of the 11-molecule cluster (B) is shown in Figure 3.12.

[c] The structure of the 9-molecule cluster (C) is shown in Figure 3.12.

[§] This value is higher by 1.18 kcal/mol than the barrier calculated for the isolated molecules in Chapter 2. The discrepancy can be explained from several points: the barrier here is for the rigid rotation; the molecular structure is from the X-ray diffraction measurements, which has some differences with the computationally optimized structure; and the computational levels are not exactly same.

The crystal of **1,4-TTB** packs in a different pattern, compared with the crystals of **2,6-TTN** or **3-TC**. Besides the close contact between the neighboring *t*-butyl groups and the central *t*-butyl group, there is also a short contact between the central *t*-butyl group and the benzene ring of neighbor 15 (benzene ring attached with *t*-butyl groups 15 and 16 in Figure 3.3). Calculations were also done to distinguish the contributions from the *t*-butyl and *t*-butyl interaction and the *t*-butyl and benzene interaction in the clusters. To evaluate the contribution from the neighboring *t*-butyl groups, they are removed in cluster 7, and hydrogen atoms are put in their place to compensate the vacant valence. To study the effect of the interaction between the benzene ring and the central *t*-butyl group, the benzene ring is removed in cluster 8 with the vacant valence compensated by hydrogen atoms. The results show that the interaction between the central *t*-butyl group and the benzene ring contributes about 40% to the rotation barrier of the central *t*-butyl group in the cluster of **1,4-TTB**.

3.5. Discussion

3.5.1. Packing Pattern of the *t*-Butyl Groups in the Crystalline Solids

It is helpful to understand the intermolecular interactions by reviewing the crystal packing structures before discussing the rotational barriers in the crystalline solids. There is only one unique molecular site in all the three crystals studied here. In Figure 3.14, the representative structures are shown. There is a similarity between the crystals of **3-TC** and **2,6-TTN**, but the crystal of **1,4-TTB** is quite different. In the crystal structures of **3-TC** and **2,6-TTN**, the four molecules in the unit cell of **3-TC** and the two molecules in the unit cell of **2,6-TTN** are nearly parallel, if the chrysene ring and the naphthalene ring are considered as a cylinder along the long axis direction of the molecules. But in the crystal of **1,4-TTB**, the two molecules in the unit cell are nearly perpendicular to each other. (Actually, there is another polymorph of the crystal **2,6-TTN** with the entry code of KOKQUW10 [39] in CCDC which shows the similar packing pattern as in **1,4-TTB**. This polymorph is not studied here.)

In the crystal packing pattern of **3-TC** and **2,6-TTN**, the aromatic rings form layer-by-layer pattern, and the *t*-butyl groups can be viewed as protruding out at the two ends of the aromatic ring layers. Besides there is no close contact between the *t*-butyl groups and the neighboring aromatic rings, the layers of *t*-butyl groups are also isolated by the aromatic rings. Each layer of the *t*-butyl groups can be considered in a pseudo 2-dimenisonal interaction network of the *t*-butyl groups. However, this view can not be applied to the crystal of **1,4-TTB** due to two reasons. The first reason is that each of the *t*-butyl groups has a close benzene ring neighbor in which the interaction between the two is significant. (In fact, the center of the close-contact neighboring benzene ring is right above the nearly in-plane methyl group of the *t*-butyl group with a distance of 3.7 Å from the ring center to the Cm atom.) The second reason is that due to the short length of the benzene ring, the interaction between the two neighboring layers of *t*-butyl groups is also very strong. One layer of the *t*-butyl groups can not be isolated from nearby layers. The 3-dimensional intermolecular interaction network must be considered.

To further demonstrate the 2-dimensional *t*-butyl layer structure, we can project one of the layers on a plane parallel to the *t*-butyl layers. This kind of projection view is shown in

Figure 3.15 for the 13-molecule cluster of **3-TC** and the 11-molecule cluster of **2,6-TTN**. Each *t*-butyl group in Figure 3.15 is presented in a triplet of colored disks. The radius of each disk approximates the van der Waals radius a methyl group, 2.02 Å [68]. The color stands for the relative height of the *t*-butyl groups. The yellow ones are under the paper (the projection plane), and they are the up-pointing *t*-butyl groups from the molecules below the plane. The pink ones are above the paper, and they are the downward *t*-butyl groups from the molecules higher than the plane. The projection plane is chosen to be equidistant to the planes formed by the yellow *t*-butyl groups and the pink *t*-butyl groups. The color intensity of each disk stands for the distance to the projection plane. The darker color is closer to the plane. For **3-TC**, the darker pink or yellow represents a distance of 0.167 Å from the methyl carbon Cm to the plane while the two lighter colors stand for a distance of 1.050 and 1.257 Å. For **2,6-TTN**, the lighter color stands for a distance of 2.602 Å, and the thicker colors stand for a distance of 1.489 and 1.436 Å.

From the graphs, we see the central *t*-butyl group is surrounded by a few close-by neighboring *t*-butyl groups. To identify how close the *t*-butyl groups are to the central *t*-butyl group, we can use the distance between the center carbons (Ct) of the two *t*-butyl groups as an index. These distances are shown in Figures 3.16, 3.17 and 3.18.

However, the interaction between the two *t*-butyl groups depends not only on the distance between the two Ct atoms, but also on their orientations. A more accurate model should evaluate the nine interactions between the methyl groups on the two *t*-butyl groups. The interaction between methane dimers has been studied by many authors using electron correlation models and density functional theory [68-71]. The Lennard-Jones intermolecular potential is in the form, $E = \frac{a}{r^{12}} - \frac{b}{r^{6}}$ [68]. We may approximate the inter-methyl interaction in the molecular crystals using this potential form. A rough estimate of the interaction between two *t*-butyl groups can be quickly obtained by summing up the nine inter-methyl interactions. In this approximation scheme, the essential information required is the distances between the 9-pairs of methyl groups on the two *t*-butyl groups.

We created Figures 3.16, 3.17 and 3.18 to show how the inter-methyl distances change with respect to the orientations of the *t*-butyl groups. In each figure, several close-contact *t*-butyl neighbors are shown, and the rest of the *t*-butyl groups are too far away and have no effect on the central *t*-butyl rotation. Above each graph in each figure, the index of the inter-Ct distance is labeled. Each graph consists of 9-sub-graphs, representing the 9 inter-methyl carbon distances between the neighbor *t*-butyl group and the central *t*-butyl group. The horizontal axis α of each sub-graph ranges from 0 to 120°, standing for the rotational angle of the central *t*-butyl group from its thermal equilibrium orientation. The vertical axis β, ranging from 0 to 120°, is the rotational angle of the neighboring *t*-butyl group from its thermal equilibrium orientation. The distance at each point on the graphs is represented by its color, as shown by the legend. The van der Waals equilibrium distance between the two methyl groups is about 4.04 Å (approximation in the equilibrium distance of the methane dimer.), which is in green color. The longer distance, in green to yellow, is in the region of the long range attractive interaction. The shorter distance, in the color from blue to red, is where the repulsive interaction plays a role. Since the repulsive interaction in the short to intermediate range is much stronger than the attractive interaction in the long range, we focus on the repulsive interaction range only.

For **3-TC**, there are two neighbors (7 and 5), showing the repulsive hot spots, which are in blue to red. The central *t*-butyl group is likely to avoid passing through the hot spots. For **2,6-TTN**, there are four neighbors with the strong repulsive interactions to the central *t*-butyl group. Particularly, for the closest neighbor (*t*-butyl 22), in term of the inter-Ct distance, the inter methyl carbon distances are constant, *i.e.* independent of their relative orientations. Therefore, we should expect there is little contribution to the central *t*-butyl rotation from this neighbor. This is confirmed by the intermolecular interaction analysis, as shown in Table 3.9.

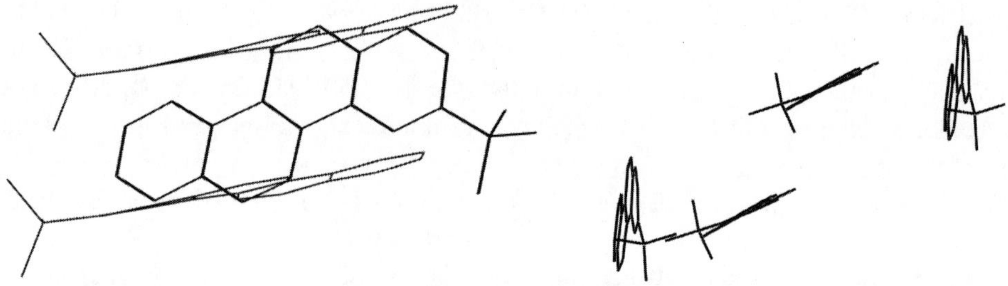

3-*t*-butylchrysene

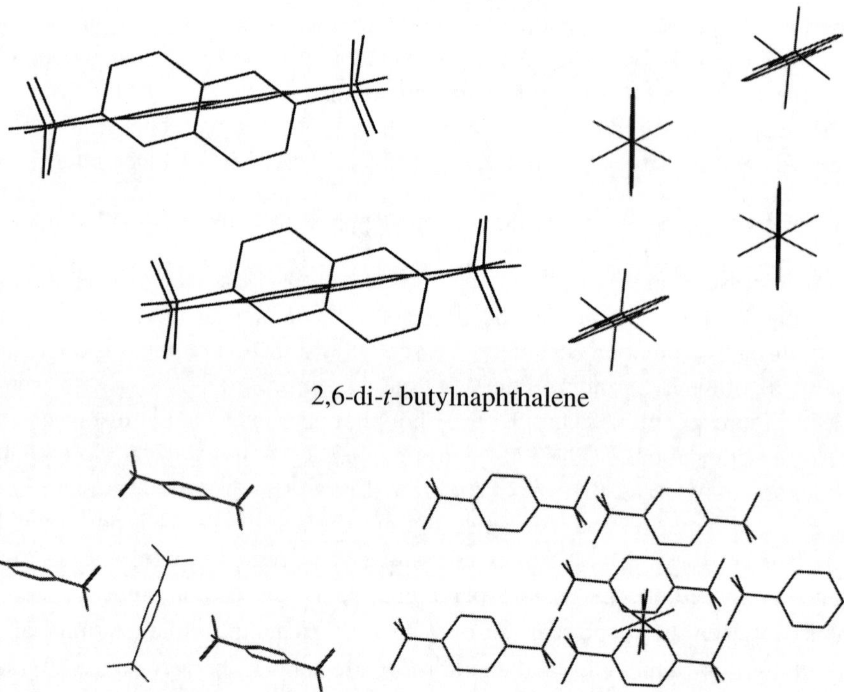

2,6-di-*t*-butylnaphthalene

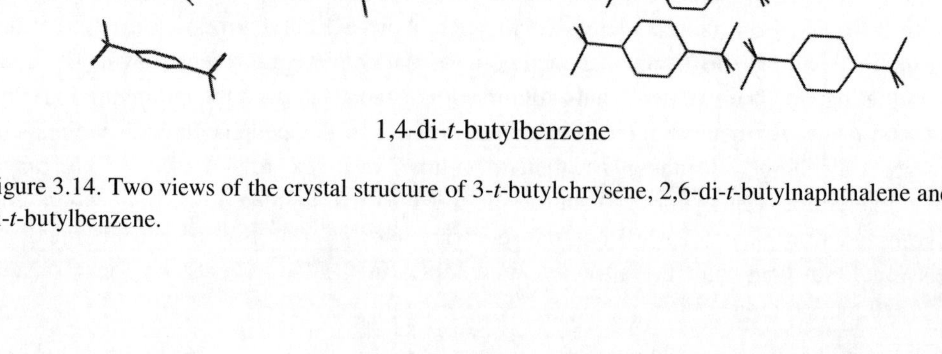

1,4-di-*t*-butylbenzene

Figure 3.14. Two views of the crystal structure of 3-*t*-butylchrysene, 2,6-di-*t*-butylnaphthalene and 1,4-di-*t*-butylbenzene.

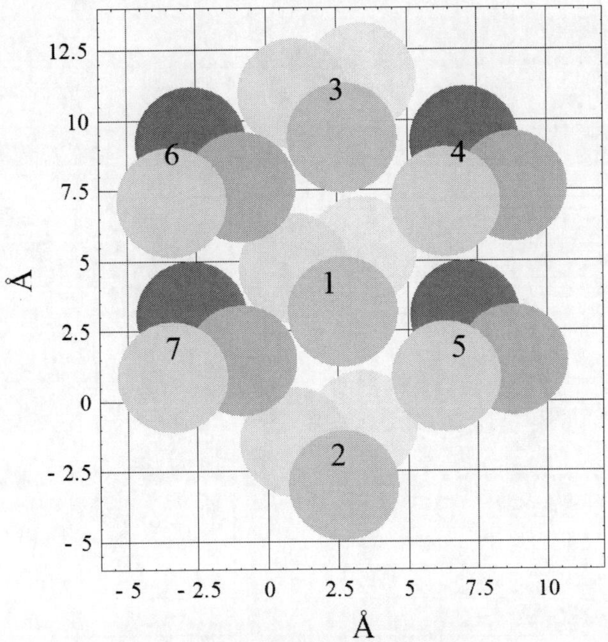

3-*t*-butylchrysene

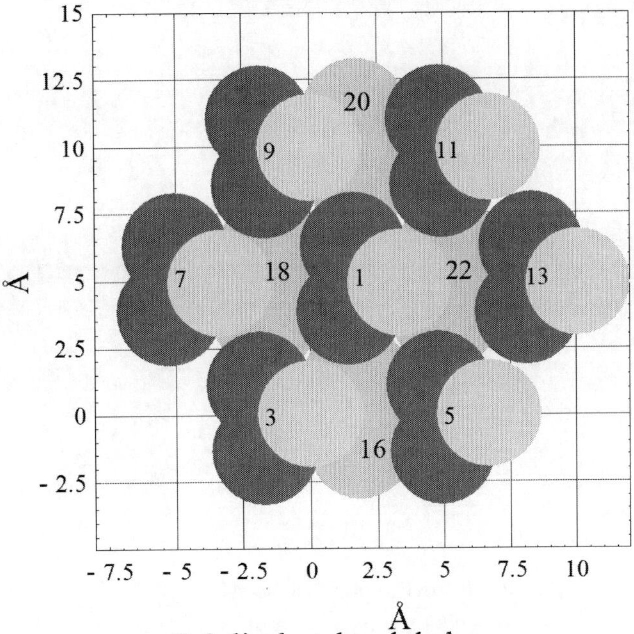

2,6-di-*t*-butylnaphthalene

Figure 3.15. Projection of the packing pattern in the clusters of 3-*t*-butylchrysene and 2,6-di-*t*-butylnaphthalene. The *t*-butyl groups in yellow are below the paper plane and the pink ones are above the plane. The diameter of each methyl disk is 4.040 Å. The deep color (pink or yellow) in **3-TC** stands for distance of 0.167 Å and the two light colors stand for the distance of 1.050 and 1.257 Å .The light color (pink or yellow) in **2,6-TTN** stands for the distance of 2.602 Å and the two deeper colors stand for the distance of 1.489 and 1.436 Å.

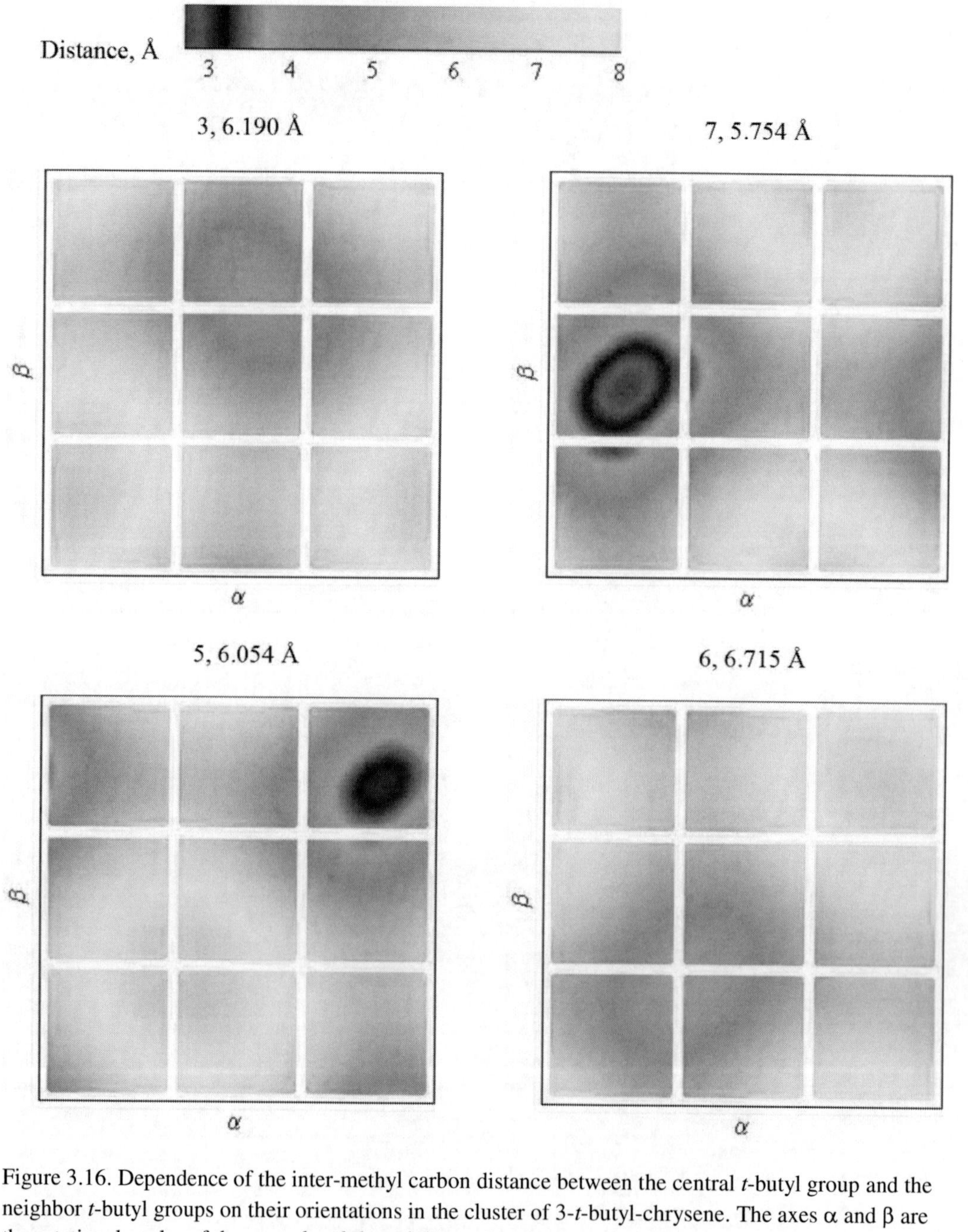

Figure 3.16. Dependence of the inter-methyl carbon distance between the central *t*-butyl group and the neighbor *t*-butyl groups on their orientations in the cluster of 3-*t*-butyl-chrysene. The axes α and β are the rotational angles of the central and the neighbor *t*-butyl groups from their equilibrium orientation, ranging from 0 to 120° for each sub-graph. The neighbor *t*-butyl index and the inter-Ct distance are shown above each graph.

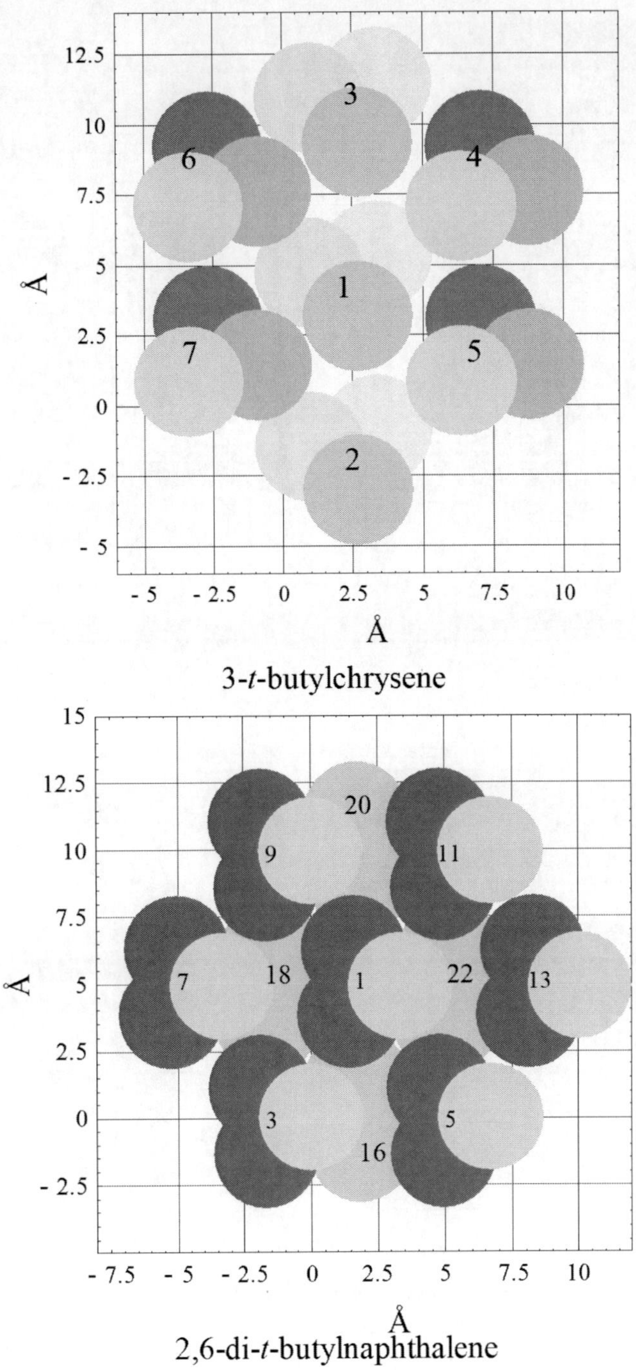

Figure 3.15. Projection of the packing pattern in the clusters of 3-*t*-butylchrysene and 2,6-di-*t*-butylnaphthalene. The *t*-butyl groups in yellow are below the paper plane and the pink ones are above the plane. The diameter of each methyl disk is 4.040 Å. The deep color (pink or yellow) in **3-TC** stands for distance of 0.167 Å and the two light colors stand for the distance of 1.050 and 1.257 Å .The light color (pink or yellow) in **2,6-TTN** stands for the distance of 2.602 Å and the two deeper colors stand for the distance of 1.489 and 1.436 Å.

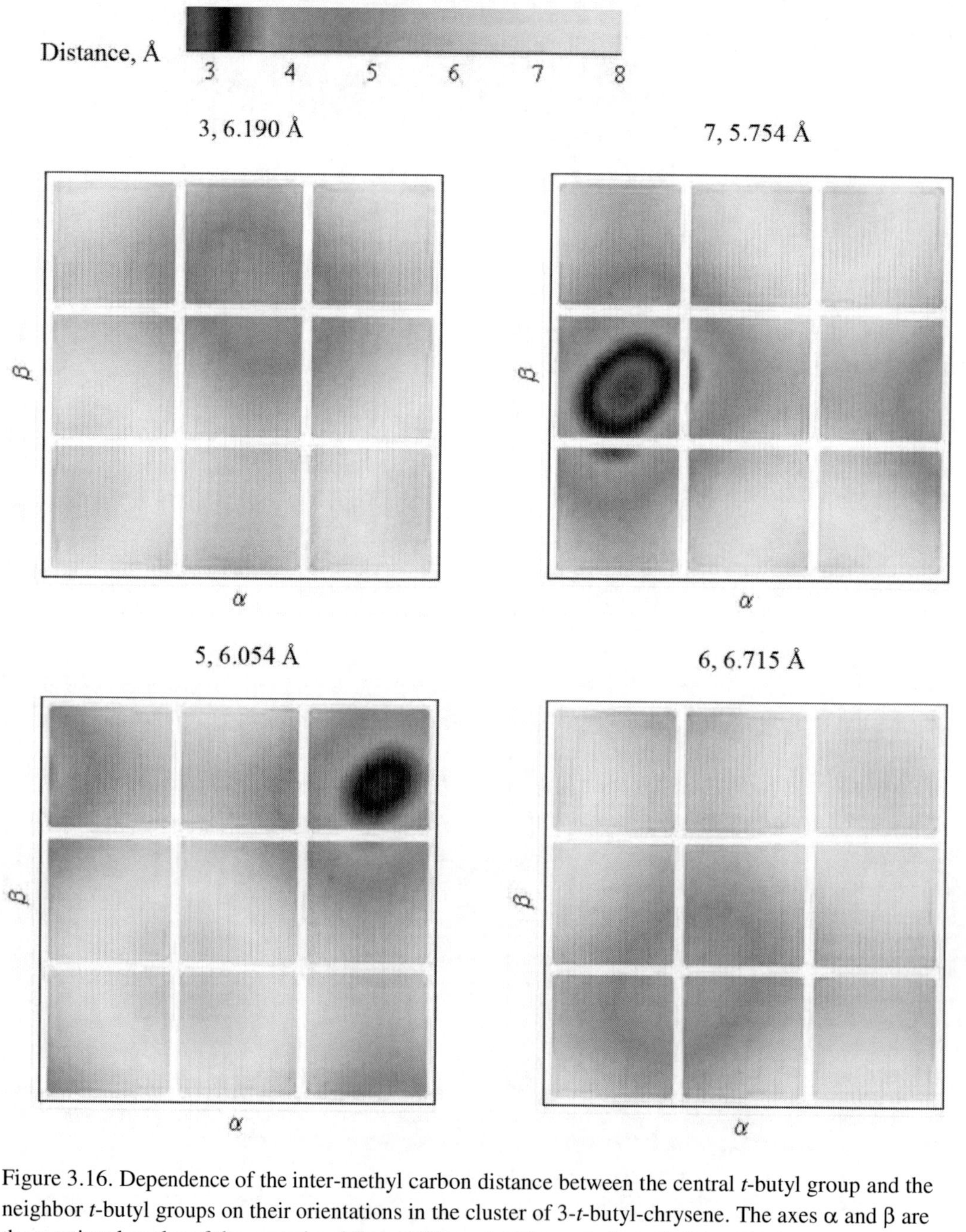

Figure 3.16. Dependence of the inter-methyl carbon distance between the central *t*-butyl group and the neighbor *t*-butyl groups on their orientations in the cluster of 3-*t*-butyl-chrysene. The axes α and β are the rotational angles of the central and the neighbor *t*-butyl groups from their equilibrium orientation, ranging from 0 to 120° for each sub-graph. The neighbor *t*-butyl index and the inter-Ct distance are shown above each graph.

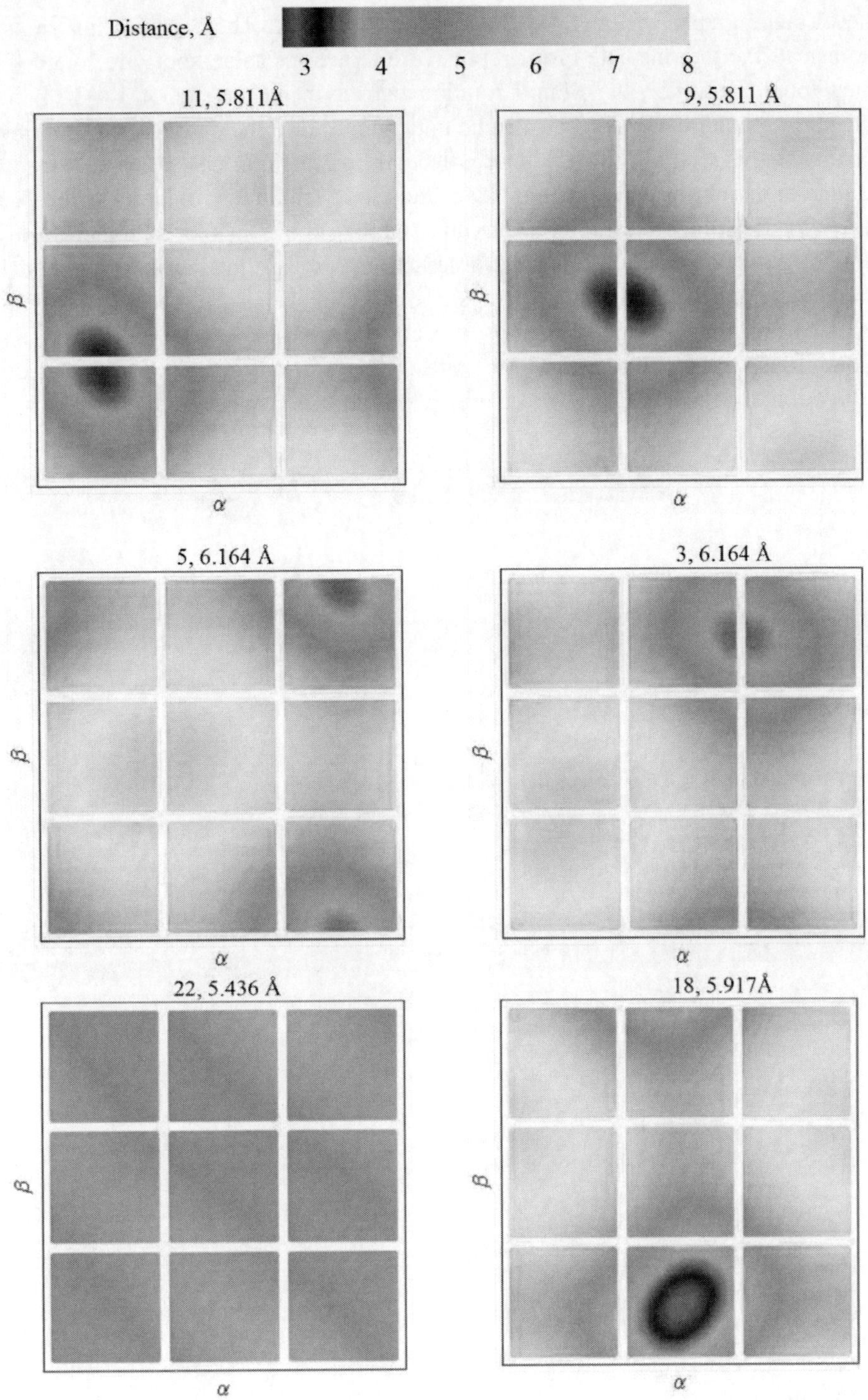

Figure 3.17. Dependence of the inter-methyl carbon distance between the central and the neighbor *t*-butyl groups on their orientations in the cluster of 2,6-di-*t*-butylnaphthalene . The axes α and β are the rotational angles of the central and the neighbor *t*-butyl groups from their equilibrium orientation, ranging from 0 to 120° for each sub-graph. The neighbor *t*-butyl index and the inter-Ct distance are shown above each graph.

The distance graphs of the **1,4-TTB** are quite different. There are no hot spots on the graphs; instead the intermediate range repulsive interactions (blue spot) are found from 8 *t*-butyl neighbors. However, the *t*-butyl rotation barrier in the crystal of **1,4-TTB** has been found smallest among the three. This can be understood from the dual roles of intermolecular interactions in the rotation barrier. Intermolecular interactions can stabilize the rotational ground state or destabilize the transition state, and thus push up the rotation barrier. It can also be a factor to destabilize the ground state and stabilize the transition state and to lower the barrier. In many situations, intermolecular interactions increase the barrier through stabilizing the ground state and destabilizing the transition state. If the intermolecular interactions are diverse and from many spatial directions, the collective result is more likely to cancel out their destabilizing effects to the transition state. If the intermolecular interaction is dominant in one direction, the result is more deterministic to favor the ground state.

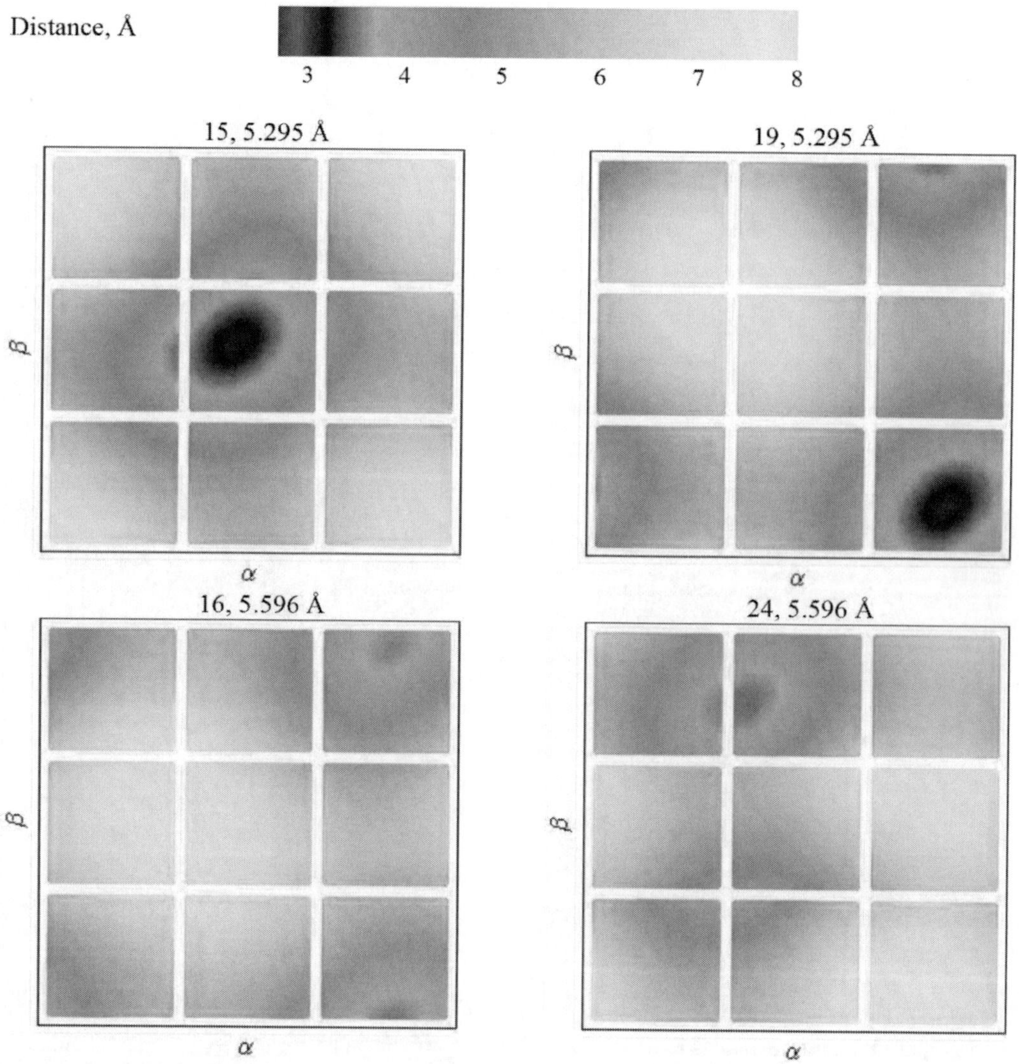

Figure 3.18. (Continued).

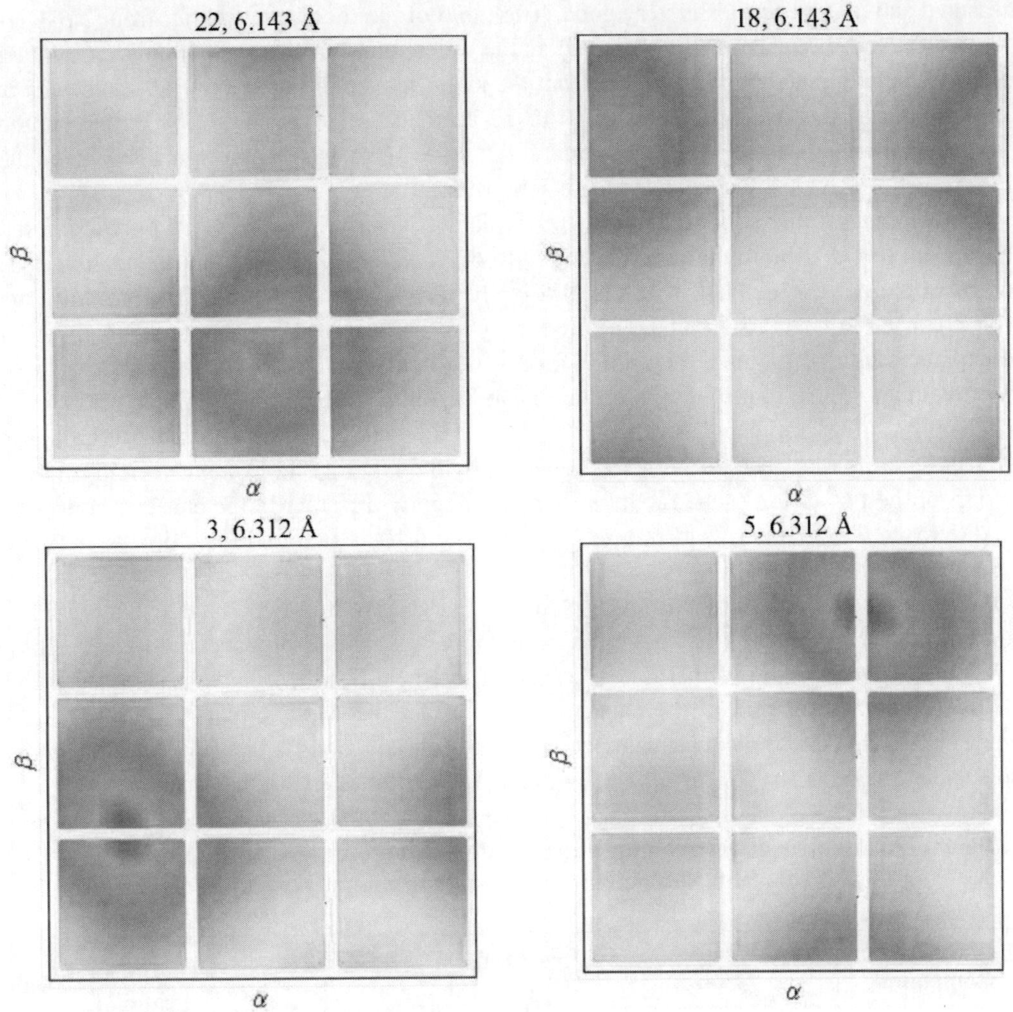

Figure 3.18. Dependence of the inter-methyl carbon distance between the central and the neighbor *t*-butyl groups on their orientations in the cluster of 1,4-di-*t*-butylbenzene . The axes α and β are the rotational angles of the central and the neighbor *t*-butyl groups from their equilibrium orientation, ranging from 0 to 120° for each sub-graph. The neighbor *t*-butyl index and the inter-Ct distance are shown above each graph.

3.5.2. Barriers to the Internal Rotations of the *t*-Butyl and Methyl Groups in the Crystalline Solids

In Table 3.11, the calculated barriers in the clusters are compared with the reorientation activation energies measured in the polycrystalline samples by solid state NMR relaxometry [38-41]. The computed barriers to the rigid rotation of the *t*-butyl groups are found to be about 1.0 kcal/mol higher than the measured values. If the *t*-butyl groups and/or the aromatic rings in the clusters are allowed to relax during the internal rotation, the computed barriers are lowered by 0.2 to 0.4 kcal/mol. The agreement between the computed barriers and the measured activation energies is very good. The trend of the increasing barrier from **2,6-TTN** to **1,4-TTB** and then to **3-TC** is the same as the order found in the experiments. The larger values of the computed barriers suggest that the slight movement of the overall molecules in the crystals and the structural relaxation of the aromatic systems could also influence the internal rotation process in the crystalline solid state. If such relaxation is allowed in the clusters, better agreement is expected to be achieved.

The barriers to the rigid rotation of the in-plane or nearly in-plane methyl groups in the clusters are found to be in the range from 5.3 to 6.1 kcal/mol. The structural relaxation does not affect the barriers too much. In the NMR relaxometry, the in-plane methyl groups have been proposed to have a same hopping frequency as the overall *t*-butyl groups, which could be explained by that the in-plane methyl groups re-orient simultaneously with the overall *t*-butyl rotations. This point is discussed later. The two out-of-plane methyl groups may have different local environments in the crystalline solid state even if the *t*-butyl group is symmetric to the ring. The difference is reflected by the fact that the computed rigid rotation barriers of the two out-of-plane methyl groups are slightly different in the clusters. However, the difference is very small. It is impossible to distinguish the two rotational processes in the experiments.

The computed barriers of the out-of-plane methyl groups are in good agreement with the experimental values of the reorientation activation energy. The discrepancy between the two is less than 0.3 kcal/mol. Furthermore, the good agreement between the values calculated in the isolated molecules and in the clusters indicates that the methyl rotation involves local interactions. The intermolecular interactions, and consequentially the shift of the whole molecule or the relaxation of the aromatic systems, have little effect to the rotation.

Table 3.11. Comparison between the calculated rotation barriers in the clusters and the measured reorientation activation energies in the polycrystalline solids

Compound	Group	B3LYP/6-31G(d), kcal/mol		Experimental, kcal/mol[b]
		Rigid	Partially Relaxed	
3-*t*-butyl-chrysene	*t*-butyl	6.66	6.26	5.7 ± 0.2[41]
	in-plane methyl	5.27		
	out-of-plane methyl	3.87(3.50)		3.4 ± 0.2
2,6-di-*t*-butyl-naphthalene	*t*-butyl	5.32	5.09	4.3 ± 0.2[39-40]
	in-plane methyl	5.99		
	out-of-plane methyl	4.04		4.3 ± 0.2
1,4-di-*t*-butyl-benzene	*t*-butyl	5.64	5.24	4.5 ± 0.2[38]
	in-plane methyl	6.07[a]		
	out-of-plane methyl	5.22(5.19)[a]		3.8 ± 0.2

[a] Calculated at the HF/3-21G level.

[b] NMR relaxometry measurement.

3.5.3. Contributions of the Intramolecular and Intermolecular Interactions to the Barriers

The comparison between the results in the isolated molecules and in the clusters provides a window to investigate the role of the *intra*molecular and *inter*molecular interactions in the internal rotation processes. Comparing the barriers to the internal rotation of the *t*-butyl groups in the two phases, the barriers in the isolated molecules are much smaller than those in the clusters and the experimental values. If we assume a same contribution from the *intra*molecular interactions in the clusters as in the isolated molecules, which is 1.14 kcal/mol in **3-TC**, 1.30 kcal/mol in **2,6-TTN** and 0.68 kcal/mol in **1,4-TTB**, 75% to 90% of the barriers come from the *inter*molecular interactions. The same conclusion can also be drawn from the intermolecular interaction analysis using the rigid *t*-butyl rotation barrier in the reduced-size clusters. (See Tables 3.8, 3.9 and 3.10.) However, one may notice that the rigid rotation barrier on a single molecule coming from the crystals is higher by about 1.0 kcal/mol than the fully relaxed rotation barrier calculated in the isolated molecules. The difference can be explained from several aspects: (1) The rigid barrier does not allow the structural relaxation; (2) The rigid barrier is calculated on the molecular structure determined by X-ray diffraction, which is slightly different from the computed structure; (3) The rigid rotation barrier is calculated at the B3LYP/6-31G(d) level while the barrier is calculated at the HF/6-311+G(d,p)// HF/6-31G(d) level in the isolated molecules. In addition, it might be speculated that the structural relaxation may be affected by the neighbors in the crystals and the molecular structural difference exists between the crystal structure and the structure of the isolated molecule. These two factors are capable of affecting the contribution of the intramolecular interaction in the *t*-butyl rotation barrier physically. But the effect should be small as shown by the two calculation schemes.

The *t*-butyl rotation PES can be decomposed into the three-fold and six-fold components. As discussed in the previous chapter, the intramolecular contribution of the six-fold component is mainly from the intramolecular steric effect while the three-fold component is proportional to the bond order difference of the two aromatic ring bonds adjacent to the *t*-butyl group. In the clusters, there is a slight increase on the six-fold component but a huge increase of the three-fold component. See Table 3.2. The increase of the three-fold component has its root in the intermolecular interactions in the clusters. The high three-fold component causes the intermediate state (τ(*t*-butyl) = 60°) in the isolated molecules to become the transition state in the clusters. Interestingly, the contribution from the intermolecular interactions has the same three-fold cosine form as the contribution from the intramolecular interactions in the clusters of **3-TC** and **2,6-TTN**. In the clusters of **3-TC** and **2,6-TTN**, the three-fold components from the intramolecular and the intermolecular interactions have the same minimum point at τ(*t*-butyl) = 0°. But in the cluster of **1,4-TTB**, the ground state orientation of the *t*-butyl groups shifts by 8.3° from the 0° in the isolated molecule. See Figures 3.4 and 3.5. This shifting is the collective result of the intramolecular and intermolecular interactions. Again, we see the contribution from the intermolecular interaction to the barrier is the dominant factor to the *t*-butyl rotation in the cluster. The uniqueness of **1,4-TTB** is probably due to its different packing pattern as discussed above. There is a perpendicular neighbor showing the close contact between its benzene ring and the neighboring *t*-butyl groups.

The detailed analysis of the contribution of the intermolecular interactions from each neighbor has been done in the reduced-size clusters. See Tables 3.8, 3.9 and 3.10. In the cluster of **3-TC**, four neighbors make significant contributions, and each contributes from 10% to 20% to the overall barrier. In the cluster of **2,6-TTN**, the contributions are mainly from 6 neighbors, and in the cluster of **1,4-TTB**, the contributions are mainly from 4 neighbors. Therefore, none of the neighbors has, by itself, a dominant effect on the *t*-butyl rotation. Instead, the barriers are a collective result of the intermolecular interactions from a few neighbors.

The intermolecular interactions are less influential on the rotation of the methyl groups. The methyl rotation barriers in the clusters are very close to the values in the isolated molecules, though there is a slight dependence on the crystal environment. In the isolated molecules, the PES can be fitted by a single term of a three-fold cosine potential function as discussed in the previous chapter. The rotation of the methyl groups is only dependent on the local environment, and the size of the aromatic system has little influence. In the clusters, there is a need of six-fold cosine potential term to fit the PES. As shown in Tables 3.3 and 3.4, the six-fold potential components are mostly negative. (This six-fold component can be explained as an artificial feature associated with the rigid rotation scheme though it is a possible result of the intermolecular interactions.) For one of the out-of-plane methyl groups in **3-TC**, the six-fold barrier is nearly 0 but the three-fold barrier is lowered from 4.00 kcal/mol in the isolated molecules to 3.87 kcal/mol in the clusters. (The barrier of the other out-of-plane methyl group is 3.50 kcal/mol.) The barrier is close to the measured value. In the crystal of the **3-TC**, the intermolecular interactions contribute a negative three-fold component to the PES. But in the clusters of **2,6-TTN** and **1,4-TTB**, the three-fold components are nearly the same as in the isolated molecules, 4.00 kcal/mol. There is a small negative six-fold component (about -0.5 kcal/mol) for the out-of-plane methyl groups in **2,6-TTN** and one of the out-of-plane methyl groups in **1,4-TTB**. The computed barriers are also close to the experimental values. The three-fold components of the in-plane methyl rotation in the clusters are nearly identical in the three compounds, 6.00 kcal/mol. Structural relaxations have no big effect to the barriers of the constituent methyl groups.

3.5.4. Correlated Intramolecular Rotations of the *t*-Butyl and the Constituent Methyl Groups and the Coupled Intermolecular Relaxations

In the isolated molecules, *t*-butyl rotation is weakly coupled with a 5° reorientation of two methyl groups, and the out-of-plane methyl rotation is also coupled with a 5° reorientation of the *t*-butyl group. More importantly, the rotation of the in-plane methyl group is strongly coupled with the rotation of the *t*-butyl group due to the steric repulsion. This coupling is energetically advantageous due to the low barrier to the *t*-butyl rotation. In the crystalline solids, the barrier to *t*-butyl rotation increases significantly due to the intermolecular interactions. Is the coupling motion still the minimum energy path? This question is best answered by the two-dimensional PES, as shown in Figure 3.10 in the 7-molecule cluster of **3-TC**. In the cluster, the barrier to the *t*-butyl rotation increases to about 6.0 kcal/mol, close to the constrained rotation barrier of the in-plane methyl group. The high *t*-butyl rotation barrier isolates the minimum energy paths for the in-plane and out-of-plane methyl rotations. There is no energetically favorable route which can couple the methyl rotation with the *t*-butyl rotation. The three paths for the in-plane and out-of-plane methyl groups and the *t*-butyl group are independent of each other, and the methyl rotation paths are orthogonal to the *t*-butyl rotation path, although there are possible small variations of the orientations of the groups.

Is there the possibility of the geared *t*-butyl group rotations in the crystalline solids? As seen the previous discussion, the intermolecular interactions contribute the largest part of the *t*-butyl rotation PESs. However, there is no dominant interaction from a single neighbor that is strong enough to drive two *t*-butyl groups to rotate together. Instead, the barrier is a result of the collective interactions. But on the other hand, the partially relaxed calculation shows the simultaneous reorientation of two close neighboring *t*-butyl groups in the cluster of **3-TC**, although the geared rotation is not predicted to occur. The two neighboring *t*-butyl groups can adjust their orientations by 7° to accommodate the central *t*-butyl group rotation. This 7° relaxation is well within the range of that obtained from thermal motion of the *t*-butyl groups. In the clusters of **1,4-TTB**, it is found that it is not possible for the benzene ring to make a complete rotation, which is locked by the two neighboring *t*-butyl groups above and below the ring, but a 10° relaxation motion is likely to happen.

3.6. CONCLUSION

Internal rotations of the *t*-butyl and the constituent methyl groups are studied in this chapter using the quantum mechanical electronic structure calculations. Combining the X-ray diffraction structures with the computation results, the cluster models can represent a good simulation of the environment in the crystalline solids for the *t*-butyl groups. The calculation results at the B3LYP/6-31G(d) level show a good agreement with the experimental values obtained from NMR relaxation studies. The following conclusions can be drawn from the above results and discussion.

(1) Combination of the X-ray diffraction technique and the quantum mechanical electronic calculation proves a viable tool to study barrier to the internal rotation in the crystalline solids. Theoretical modeling at the B3LYP/6-31G(d) level is capable of predicting an accurate barrier in the scale of several kcal/mol.
(2) The barrier to *t*-butyl rotation in the crystalline solid state is affected strongly by the intermolecular interactions. The calculated barriers to the *t*-butyl rotations are 6.26, 5.09 and 5.24 kcal/mol in the clusters of **3-TC**, **2,6-TTN** and **1,4-TTB**, respectively.
(3) The collective interactions from a few neighbors contribute about 70% to 90% of the *t*-butyl rotation barrier in the clusters. In the clusters of **3-TC** and **2,6-TTN**, the interactions that contribute to the barriers are mainly between the neighboring *t*-butyl groups. In the cluster of **1,4-TTB**, the interaction between the *t*-butyl group and the neighboring benzene ring also makes a significant contribution.
(4) The intermolecular interactions also affect the barriers to the internal rotation of the constituent methyl groups, but to a very small degree. The methyl rotation is mainly determined by the intramolecular interactions.
(5) Different from the isolated molecules, the methyl and *t*-butyl rotations are independent in the crystalline solids of the three compounds. The simultaneously

rotation of the in-plane methyl group and the *t*-butyl group is an energetically unfavorable route. However, the 5° reorientation of the constituent methyl groups is coupled with the *t*-butyl group rotation.

(6) The geared rotation of the *t*-butyl groups across neighbors is not found in the clusters. But the simultaneous reorientation in 7° or smaller of a few close-by neighboring *t*-butyl groups is found to accompany the *t*-butyl rotation. In the cluster of **1,4-TTB**, it is also allowed for the benzene ring to adjust its orientation slightly in the *t*-butyl rotation process.

3.7. REFERENCES

[1] Mason, EA; Kreevoy, MM. A simple model for barriers to internal rotation. *J. Am. Chem. Soc.*, 1955, 77, 5808-5814.

[2] Allen, LC. Energy component analysis of rotational barriers. *Chem. Phys. Lett.*, 1968, 2(8).

[3] Radom, L; Lathan, WA; Hehre, WJ; Pople, JA. Molecular orbital theory of the electronic structure of organic compounds. XVII. Internal rotation in 1,2-disubstituted ethanes. *J. Am. Chem. Soc.*, 1973, 95(3), 693-698.

[4] Schaefer, T; Sebastian, R; Penner, GH. Theoretical and experimental data on the internal rotational potential in isopropylbenzene. *Can. J. Chem.*, 1987, 66, 1495-1499.

[5] Brunelle, JA; Ketendre, LJ; Weltin, EE; Brown, JH; Bushweller, CH. Barriers to internal rotation in CF_3CFCl_2, CF_3CHCl_2, CF_3CF_2Br, CF_3CF_2Cl, CF_3CH_2Br and CF_3CH_2Cl. F-19 and H-1 dynamic NMR studies. *Ab initio* molecular orbital barrier calculations. *J. Phys. Chem.* 1992, 96, 9225-9232.

[6] Sztraka, L; Csonka, GI. Flexible *ab initio* geometry of methylamine and its internal rotation. *J. Mol. Struct* 1997, 410-411, 387-390.

[7] Benassi, R; Taddei, FA. Theoretical approach to the factorization of the effects governing the barrier for internal rotation around the C(sp2)-C(sp3) bond into α-substituted toluenes. *J. Mol. Struct. (Theochem)*, 1997, 418, 59-71.

[8] Yanborisov, VM; Zaitesev, MY; Monakov, YB. Determination of internal rotation barrier. *Russian J. App. Chem.*, 2001, 74(12), 1942-1944.

[9] Cacelli, I; Prampolini, G. Torsional barriers and correlations between dihedarls in p-polyphenyls. *J. Phys. Chem. A*, 2003, *107*, 8665-8670.

[10] Palmer, MH; Camp, PJ. An *ab initio* study of vibration and rotation effects in isocyanatometane, isothiocynantomethane and azidomethane. *Mol. Phys.*, 2003, 101(20), 3053-3062.

[11] Halpern, AM; Glendening, ED. An intrinsic reaction coordinate calculation of the torsional potential in ethane: comparison of the computationally and experimentally derived torsional transitions and the rotational barrier. *J. Chem. Phys.*, 2003, 119(21), 11186-11191.

[12] Zierkiewicz, W; Michalska, D; Hobza, P. The barrier to internal rotation and electronic effects in *para*-halogenophenols: Theoretical study. *Chem. Phys. Lett.*, 2004, 386, 95-100.

[13] Stevenson, PE; Bhat, G; Bushweller, CH; Anderson, WG. Experimental and theoretical studies of the rate of *tert*-butyl rotation in 2-*tert*-butyl-1,3-diheteroatomic rings. The *gem*-dialkyl effect, ring geometry, and lone pair orientation as factors in conformational dynamics. *J. Am. Chem. Soc.*, 1974, 96, 1067-1071.

[14] Pacansky, J; Yoshimine, M. Theoretical studies on the barriers for internal rotation of the methyl groups in the *tert*-butyl radical. *J. Phys. Chem.*, 1986, 90, 1980-1983.

[15] Richardson, PR; Chapman, MA; Wilson, DC; Bates, SP; Jones, AC. The nature of conformational preference in a number of *p*-alkylphenols and *p*-alkyl benzenes. *Phys. Chem. Chem. Phys.*, 2002, 4, 4910-4915.

[16] Chen, C.-C; Bozzelli, JW. Structures, intramolecular rotation barriers, and thermochemical properties of methyl ethyl, methyl isopropyl, and methyl *tert*-butyl ethers and the corresponding radicals. *J. Phys. Chem. A*, 2003, 107, 4531-4546.

[17] Sonoda, Y; Iwata, S. Theoretical studies of the internal rotation of the methyl group in *o*-, *m*-, and *p*-fluorotoluenes and their cations. *Chem. Phys. Lett.*, 1995, 243(1/2), 176-182.

[18] Zierkiewicz, W; Michalska, D; Hobza, P. The barrier to internal rotation and electronic effects in *para*-halogenophenols: theoretical study. *Chem. Phys. Lett.*, 2004, 386, 95-100.

[19] Goodman, L; Pophristic, V; Weinhold, F. Origin of methyl internal rotation barriers. *Acc. Chem. Res.*, 1999, 32(12), 983-993.

[20] Pophristic, V; Goodman, L. Hyperconjugation not steric repulsion leads to the staggered structure of ethane. *Nature*, 2001, 411, 565-568.

[21] Price, SL. Toward more accurate model intermolecular potentials for organic molecules. In *Reviews in Computational Chemistry*, Lipkowitz, K. B. and Boyd, D., B., Eds. Wiley-VCH, John Wiley and Sons, Inc.: New York, 225, Vol. 14, 225-289.

[22] Stone, AJ. *The Theory of Intermolecular Forces*. Oxford: New York, 1996, Vol. 32, p 264.

[23] Makov, G; Payne, MC. Periodic boundary conditions in *ab initio* calculations. *Phys. Rev. B*, 1995, 51(7), 4014-4022.

[24] Cardenas-Jiron, G; Masunov, A; Dannenberg, JJ. Molecular orbital study of crystalline *p*-benzoquinone. *J. Phys. Chem. A*, 1999, 103, 7042-7046.

[25] Kudin, K; Scuseria, GE. Linear-scaling density-functional theory with gaussian orbitals and periodic boundary conditions: Efficient evaluation of energy and forces via the fast multipole method. *Phys. Rev. B*, 2000, 61(24), 16440-16453.

[26] Ouvrard, C; Price, SL. Toward crystal structure prediction for conformationally flexible molecules: the headaches illustrated by aspirin. *Crystal Growth & Design*, 2004, 4(6), 1119-1127.

[27] Dovesi, R; Civalleri, B; Orlando, R; Roetti, C; Saunders, VR. *Ab initio* quantum simulation in solid state chemistry. In *Reviews in Computational Chemistry*, Lipkowitz, K. B., Larter, R. and Cundari, T. R., Eds. Wiley-VCH, John Wiley & Sons, Inc.: 2005, Vol. 21, pp 1-125.

[28] Pisani, C; Busso, M; Capecchi, G; Casassa, S; Dovesi, R; Maschio, L; Zicovich-Wilson, C; Schutz, M. Local-MP2 electron correlation method for nonconducting crystals. *J. Chem. Phys.* 2005, 122(9), 094113/1-094113/12.

[29] Suhai, S; Ladik, J. *J. Phys. Chem.* 1982, 15, 4327.

[30] Suhai, S. *Phys. Rev. B*, 1983, 27, 3506.

[31] Sun, J.-Q; Bartlett, RJ. *J. Chem. Phys.* 1996, 104, 8553.
[32] Ayala, PY; Kudin, KN; Scuseria, GE. *J. Chem. Phys.*, 2001, 115, 9698.
[33] Förner, W. Chem. Phys., 1985, 97, 251.
[34] Förner, W; Knab, R; Cizek, J; Ladik, J. *J. Chem. Phys.*, 1997, 106, 10248.
[35] Lüchow, A; Anderson, JB. *Annu. Rev. Phys. Chem.*, 2000, 51, 501.
[36] Zimmerman, HE; Nesterov, EE. *J. Am. Chem. Soc.*, 2002, 124, 2818.
[37] Dillen, J; Verhoeven, P. The end of a 30-year-old controversy? A computational study of the B-N stretching frequency of BH_3-NH_3 in the solid state. *J. Phys. Chem. A*, 2003, 107, 2570-2577.
[38] Beckmann, PA; Fusco, FA; O'Neill, AE. Proton spin-lattice relaxation and intramolecular reorientation in solids. *J. Magn. Reson.*, 1984, 59, 63-70.
[39] Rheingold, AL; Figueroa, JS; Dybowski, C; Beckmann, PA. Superlattices, polymorphs and solid-state NMR spin–lattice relaxation (T_1) measurements of 2,6-di-*tert*-butylnaphthalene. *Chem. Comm.*, 2000, 651-652.
[40] Beckmann, PA; Burbank, KS; Clemo, KM; Slonaker, EN; Averill, K; Dybowski, C; Figueroa, JS; Glatfelter, A; Koch, S; Liable-Sands, LM; Rheingold, AL. ^{1}H nuclear magnetic resonance spin-lattice relaxation,^{13}C magic-angle-spinning nuclear magnetic resonance spectroscopy, differential scanning calorimetry, and X-ray diffraction of two polymorphs of 2,6-di-*tert*-butyl-naphthalene. *J. Chem. Phys.*, 2000, 113, 1958-1965.
[41] Beckmann, PA; Buser, CA; Gullifer, K; Mallory, FB; Mallory, CW; Rossi, GM; Rheingold, AL. Methyl and *t*-butyl group reorientation in planar aromatic solids: Low-frequency nuclear magnetic resonance relaxometry and X-ray diffraction. *J. Chem. Phys.*, 2003, 118(24), 11129-11138.
[42] Beckmann, PA; Paty, C; Allocco, E; Herd, M; Kuranz, C; Rheingold, AL. The relationship between crystal structure and methyl and *t*-butyl group dynamics in van der Waals organic solids. *J. Chem. Phys.*, 2004, 120(11), 5309-5314.
[43] Maseras, F; Morokuma, K., *J. Comp. Chem.*, 1995, 16, 1170.
[44] Humbel, S; Sieber, S; Morokuma, K., *J. Chem. Phys.*, 1996, 105, 1959.
[45] Matsubara, T; Sieber, S; Morokuma, K., *Int. J. Quant. Chem.* 1996, 60, 1101.
[46] Svensson, M; Humbel, S; Froese, RDJ; Matsubara, T; Sieber, S; Morokuma, K., *J. Phys. Chem.*, 1996, 100, 19357.
[47] Svensson, M; Humbel, S; Morokuma, K., *J. Chem. Phys.* 1996, 105, 3654.
[48] Dapprich, S; Komáromi, I; Byun, KS; Morokuma, K; Frisch, MJ., *J. Mol. Struct. (Theochem)*, 1999, 462, 1.
[49] Vreven, T; Morokuma, K., *J. Comp. Chem.*, 2000, 21, 1419.
[50] The X-ray crystal structures are available free of charge from the Cambridge Crystallographic Data Centre. See http://www.ccdc.cam.ac.uk/ for details.
[51] Rheingold, AL. *Private Communication*, 2001.
[52] Kravers, MA; Antipin, MYu; Struchkov, YuT. *Cryst. Struct. Commun.,* 1980, 9, 955. (BUTBNZ)
[53] Allen, FH. A systematic pairwise comparison of geometric parameters obtained by X-ray and neutron diffraction. *Acta Cryst.*, 1986, B42, 515.
[54] Allen, FH; Kennard, O; Watson, DG; Brammer, L; Orpen, AG; Taylor, R., *J. Chem. Soc. Perkin Trans.2*, 1987, S1-S19..
[55] Goodman, L; Pophristic, V. Origin of methyl internal rotation barriers. *Acc. Chem. Res.*, 1999, 32, (12), 983-993.

[56] Pophristic, V; Goodman, L. Hyperconjugation not steric repulsion leads to the staggered structure of ethane. *Nature* (London), 2001, 411, 565-568.
(56a) Hydrogen atoms are located in (difference) electron-density maps in the final stages of structure determination. A variety of H-atom refinement strategies are then employed. See http://www.ccdc.cam.ac.uk/support/documentation/conquest/ConQuest_17-1-276.html.

[57] Frisch, MJ; Trucks, GW; Schlegel, HB; Scuseria, GE; Robb, MA; Cheeseman, JR.; Montgomery, Jr. JA; Vreven, T; Kudin, KN; Burant, JC; Millam, JM; Iyengar, SS; Tomasi, J; Barone, V; Mennucci, B; Cossi, M; Scalmani, G; Rega, N; Petersson, GA; Nakatsuji, H; Hada, M; Ehara, M; Toyota, K; Fukuda, R; Hasegawa, J; Ishida, M; Nakajima, T; Honda, Y; Kitao, O; Nakai, H; Klene, M; Li, X; Knox, JE; Hratchian, HP; Cross, JB; Bakken, V; Adamo, C; Jaramillo, J; Gomperts, R; Stratmann, RE; Yazyev, O; Austin, AJ; Cammi, R; Pomelli, C; Ochterski, JW; Ayala, PY; Morokuma, K; Voth, GA; Salvador, P; Dannenberg, JJ; Zakrzewski, VG; Dapprich, S; Daniels, AD; Strain, MC; Farkas, O; Malick, DK; Rabuck, AD; Raghavachari, K; Foresman, JB; Ortiz, JV; Cui, Q; Baboul, AG; Clifford, S; Cioslowski, J; Stefanov, BB; Liu, G; Liashenko, A; Piskorz, P; Komaromi, I; Martin, RL; Fox, DJ; Keith, T; Al-Laham, MA; Peng, CY; Nanayakkara, A; Challacombe, M; Gill, PMW; Johnson, B; Chen, W; Wong, MW; Gonzalez, C; Pople, JA. *Gaussian 03, Revision B.05*, Gaussian, Inc., Pittsburgh PA, 2003.

[58] Schwenke, DW; Truhlar, DG. *J. Chem. Phys.*, 1984, *82*, 2418.

[59] Frisch, J; Del Bene, JE; Binkley, JS; Schaefer, HF, III *J. Chem. Phys.,* 1986, 2279.

[60] Szalewicz, K; Cole, SJ; Kolos, W; Bartlett, RJ., *J. Chem. Phys.*, 1988, *89*, 3662.

[61] Turi, L; Dannenberg, JJ., *J. Phys. Chem.*, 1993, 97, 2488.

[62] van Duijneveldt-van Rijdt, JGCM; van Duijneveldt, FB. *Ab Initio* Methods in Quantum Chemistry; Lawley, K. P., Ed.; John Wiley & Sons: New York, 1987, Vol. II.

[63] Gutowskil, M; van Duijneveldt-van de Rijdt, JGCM; van Duijneveldt, F. B. *J. Chem. Phys.*, 1993, 98, 4728.

[64] Cook, DB; Sordo, JA; Sordo, TL. *Int. J. Quantum Chem.*, 1993, 48, 375.

[65] van Duijneveldt, FB; van Duijneveldt-van Rijdt, JGCM; van Lenthe, JH. *Chem. Rev.*, 1994, 94, 1873.
(65a) Zhao, Y.; Pu, J.; Lynch, B. J.; Truhlar, D. G. Tests of second-generation and third-generation density functionals for thermochemical kinetics. *Phys. Chem. Chem. Phys.* 2004, *6*, 673-676.

[66] Penner, GH; Polson, JM; Stuart, C; Ferguson, G; Kaitner, B. Deuterium NMR and X-ray crystallographic studies of guest and host motions in the thiourea/ 1,4-di-*tert*-butylbenzene inclusion compound. *J. Phys. Chem.*, 1992, 96, 5121-5129.

[67] Sidhu, PS; Penner, GH; Jeffrey, KR; Zhao, B; Wang, ZL; Goh, I. Solid-state NMR study of guest molecule dynamics in 4-alkyl-*tert*-butylbenzene/ thiourea inclusion compounds. *J. Phys. Chem. B*, 1997, 101, 9087-9097.

[68] Chen, X.-R; Bai, Y.-B; Jun, Z; Xiang-Dong, Y. Intermolecular interaction potentials of the methane dimer from the local density approximation. *Phys. Rev. A*, 2004, 69, 034701.

[69] Tsuzuki, S; Uchimaru, T; Tanabe, K. Refinement of nonbonding interaction potential parameters for methane on the basis of the pair potential obtained by MP3/6-

311G(3d,3p)-level *ab initio* molecular orbital calculations: The anisotropy of H/H interaction. *J. Phys. Chem.*, 1994, 98, 1830-1833.

[70] Urata, S; Tsuzuki, S; Mikami, M; Takada, A; Uchimaru, T; Sekiya, A. Analysis of the intermolecular interaction between CH_3OCH_3, CF_3OCH_3, CF_3OCF_3, and CH_4: High level *ab initio* calculations. *J. Comput. Chem.*, 2002, 23, 1472-1479.

[71] Harbola, U; Mukamel, S. Intermolecular forces and nonbonded interactions: superoperator nonlinear time-dependent density-functional-theory response approach. *Phys. Rev. A*, 2004, 70, 052506-1-052506-15.

Chapter 4

INTERNAL ROTATIONS OF THE PHENOLS IN THE ISOLATED MOLECULES

4.1. ABSTRACT

Internal rotations of the *t*-butyl groups and their constituent methyl groups and the hydroxyl groups are studied in the isolated molecules of 2-*t*-butyl-4-methyl-phenol (**1,2,4-OTMB**) and 2,6-di-*t*-butyl-4-methylphenol (**1,2,4,6-OTMTB**). At the HF/6-311+G(d,p)//HF/6-31G(d) level, the barrier to the 2-*t*-butyl rotation in **1,2,4-OTMB** is 3.58 kcal/mol, and the barriers to the 2-*t*-butyl rotation and the 6-*t*-butyl rotation in **1,2,4,6-OTMTB** are 4.08 and 3.09 kcal/mol, respectively. In the two compounds, the barriers to the out-of-plane methyl rotations are in the range from 3.06 to 3.36 kcal/mol, and the barriers to the in-plane methyl rotations are in the range from 7.34 to 7.69 kcal/mol. The barrier to the hydroxyl rotation is 3.11 kcal/mol in **1,2,4-OTMB** and 0.46 kcal/mol in **1,2,4,6-OTMTB**. The *t*-butyl rotations are compared with those in aromatic hydrocarbons, *e.g.* 3-*t*-butyl-chrysene, 2,6-di-*t*-butylnaphthalene and 1,4-di-*t*-butylbenzene. The presence of the hydroxyl group at the ortho position in the phenols shows significant influence on the internal rotation of the *t*-butyl groups and the constituent methyl groups. The origin of this 'hydroxyl effect' is briefly discussed. Couplings among different rotational motions are discussed in terms of fully relaxed rotation potential energy surfaces and two-dimensional rotation potential energy surfaces.

4.2. INTRODUCTION

In this chapter, internal rotations of *t*-butyl groups and constituent methyl groups are studied in the isolated molecules of two phenols. Three aromatic hydrocarbons, 3-*t*-butylchrysene, 2,6-di-*t*-butylnaphthalene and 1,4-di-*t*-butyl-benzene have been studied in the previous two chapters. In these three compounds, there is no strong steric hindrance to the internal rotation of the *t*-butyl groups in the isolated molecules. There are only weak interactions from the flanking ring hydrogen atoms. The barriers are found to be very small in the isolated molecules, in the range from 0.5 to 1.0 kcal/mol. The results are consistent with experimental measurements in the gas phase and/or in the solution phase in the literature [1-4]. The origin of the barrier is mainly from the hyperconjugative interaction between the *t*-butyl group and the aromatic ring. In the crystalline solids, the intermolecular interactions are dominant, and the barriers increase to the range from 4 to 6 kcal/mol as shown by our computational results and indicated by experimental observations [5-8].

The two phenols studied here and in the following chapter are 2-*t*-butyl-4-methylphenol (**1,2,4-OTMB**) and 2,6-di-*t*-butyl-4-methylphenol (**1,2,4,6-OTMTB**). In these two compounds, the *t*-butyl groups are at the ortho position relative to the hydroxyl groups. The presence of the hydroxyl group brings significant steric hindrance to the internal rotation of the *t*-butyl groups. In addition, the strong hyperconjugation between the hydroxyl group and the benzene ring disturbs the π-electron density distribution on the ring, which in turn may further increase or decrease the *t*-butyl rotation barrier [9-10]. In this chapter, we use electronic structure calculations to explore these problems in the isolated molecules of the compounds. The *t*-butyl rotation has been studied experimentally in the crystalline solids of the compounds using the low frequency NMR relaxometry method [5-8]. We use cluster models to simulate the environment of the crystalline solids in the following chapter.

The barrier to the hydroxyl rotation on phenol and para-halogen substituted phenols has been experimentally determined in the gas phase and in the solution phase, and it is in the range from 3.36 to 3.56 kcal/mol [11-15]. Theoretical studies in the literature reproduce the barriers well and reveal that the substituent at the para position may reinforce or oppose the delocalization of the electron density of the oxygen lone pair on the hydroxyl group to the benzene ring, and thereby affect the barrier to the hydroxyl rotation.

However, the steric-hindered rotation of the hydroxyl group by an ortho substituent on a benzene ring has not been thoroughly studied. Also the rotation of *t*-butyl groups strongly steric-hindered by intramolecular substituents has not been studied very much. Yamamoto et al. [16-17] designed a molecule (shown below), in which the *t*-butyl rotation is steric-hindered by a large group on the neighboring ring. The barrier in the solution phase, determined by the dynamic NMR method, was found to be 9.6 kcal/mol.

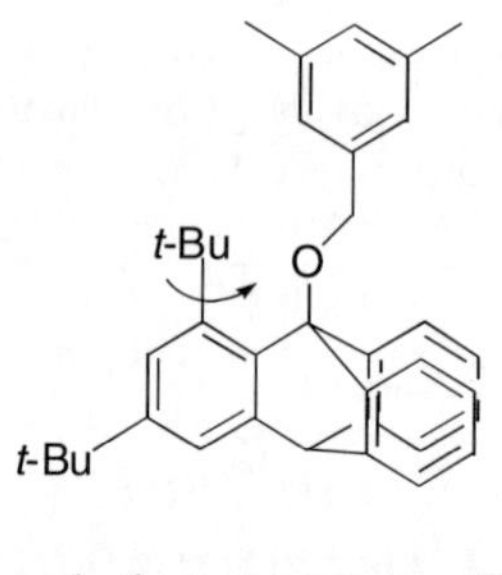

Beckmann et al. [5-8] measured the *t*-butyl reorientation activation energy in the crystalline solids for a series of phenols including **1,2,4-OTMB** and **1,2,4,6-OTMTB** using low frequency NMR relaxometry. Proton spin-lattice relaxation rates were determined as a function of the temperatures of the samples at various NMR frequencies.

There are two unique crystallographic molecular sites in the crystal of **1,2,4-OTMB**, in which the two *t*-butyl groups at the two sites have different neighborhoods. In the crystal of **1,2,4,6-OTMTB**, there is only one unique molecular site, but the two *t*-butyl groups at the 2 and 6 positions of the benzene ring are quite different in terms of both the *intra*molecular and *inter*molecular environments. The computational study provides valuable complementary information to aid the interpretation of the NMR relaxation results and to evaluate the role of intermolecular interactions in the internal rotation of the *t*-butyl groups under different environments. For different *t*-butyl groups, their local environments can be modeled in cluster models built from the crystal structure of the compound and the rotational barriers can be evaluated computationally in the clusters.

In this chapter, we present the potential energy surfaces for the different rotational degrees of freedom calculated in the isolated molecules of the phenols. We discuss how the intramolecular interactions affect the barriers. We also investigate the relationship among different degrees of freedom. We will continue our studies in the crystalline solids in the next chapter. The internal rotation of the 4-methyl group in the two compounds, which has a very low barrier, has been studied extensively both experimentally and theoretically [18-22]. The dynamics of this methyl group are observed at very low temperatures and the reorientation occurs via quantum mechanical tunneling. We are not going to look at this group here.

4.3. METHODS

All the calculations were carried out using the same procedure as in Chapter 2. See the Methods section there for details.

4.4. RESULTS

4.4.1. Labeling Scheme

To describe internal rotation coordinates and rotational conformations, we label the compounds as shown in Scheme 4.1. The carbon atoms of the benzene rings are numbered from 1 to 6 based on the IUPAC nomenclature rule. The central quaternary carbons of the *t*-butyl groups are labeled by the letter 't'. The two *t*-butyl groups on 2,6-di-*t*-butyl-4-methylphenol (**1,2,4,6-OTMTB**) can be differentiated by referring to the connected ring carbons. The constituent methyl carbons of the *t*-butyl groups are labeled by the letter 'm'. The carbon atoms of the 4-methyl groups are labeled by 'm″'. For convenience, we use the letter 'a' as a general label for the ring carbons that the *t*-butyl groups are attached to, *i.e.* C2 in **1,2,4-OTMB** and C2 and C6 in **1,2,4,6-OTMTB**. The internal rotation coordinates of the *t*-butyl groups and the hydroxyl groups are represented by one of their dihedrals relative to the benzene ring. The internal rotation coordinate of the 2-*t*-butyl group, τ(*t*-butyl), on 2-*t*-butyl-4-methylphenol (**1,2,4-OTMB**) is described by one of the dihedrals between the Cm-Ct bonds and the C2-C3 bond, τ(CmCtC2C3). The other two dihedrals are approximately equal to τ(*t*-butyl) ± 120°. In a similar definition, the internal rotation coordinates of the 2-*t*-butyl and 6-*t*-butyl groups on **1,2,4,6-OTMTB** are represented by one of the dihedrals, τ(CmCtC2C3) and τ(CmCtC6C5), respectively. The internal rotation coordinates of the hydroxyl group, τ(hydroxyl), on the two compounds are represented by the dihedral, τ(HOC1C2). In Scheme 4.1, τ(CmCtC2C3) = 0°, τ(CmCtC6C5) = 0° and τ(HOC1C2) = 180°.

The internal rotation coordinates of the constituent methyl groups, τ(methyl), are labeled by the dihedrals, τ(HCmCtC2) or τ(HCmCtC6). The conformation at τ(methyl) = 60° is a staggered conformation of the methyl group relative to the rest of the *t*-butyl group, and τ(methyl) = 0° is the eclipsed conformation. The 4-methyl group is not investigated in this study.

4.4.2. Computed Ground State Molecular Structures versus X-ray Crystal Structures

Ground state structures for the isolated molecules of **1,2,4-OTMB** and **1,2,4,6-OTMTB** were obtained by optimization at the HF/6-31G(d)//HF/6-31G(d) level. The normal mode analysis at the HF/6-31G(d) level confirmed they are global energy minimum conformations. The crystal structures of the compounds were determined by X-ray diffraction crystallography at 173 K and at room temperature, respectively [7, 25-26]. There are two unique molecular sites in the crystal of **1,2,4-OTMB** and they are shown under ‘site 1’ and ‘site 2’ in Scheme 4.3. In Tables 4.1 and 4.2, the backbone structures of the compounds are compared with the computed structures in the isolated molecules. As seen in the tables, the computed molecular structures agree well with the X-ray crystal structures except the orientations of the *t*-butyl groups.

In the computed structures of the isolated molecules, all the *t*-butyl groups prefer the conformation at τ(*t*-butyl) = 0°. In the crystal of **1,2,4-OTMB**, the molecular site 1 has τ(*t*-butyl) = -9.2° and the site 2 has τ(*t*-butyl) = -1.7°. In the crystal of **1,2,4,6-OTMTB**, the dihedrals, τ(*t*-butyl), of the 2-*t*-butyl and 6-*t*-butyl groups are 1.0° and -5.3°. (The sign of the dihedrals is meaningful in the crystals. The definition is arbitrary, but consistent throughout this thesis.) In **1,2,4,6-OTMTB**, the distinction between the 2-*t*-butyl and 6-*t*-butyl groups is arbitrary but consistent. The 6-*t*-butyl group is closer to the hydrogen atom of the hydroxyl group than the 2-*t*-butyl group.

Besides the difference in the *t*-butyl orientations, the computed C-O bonds are slightly shorter than the values found in the X-ray crystal structure by 0.013 and 0.028 Å. The hydrogen positions are not available from the X-ray crystal structures. They are obtained by computation in the cluster models, as discussed in the following chapter. In the computed structures of the isolated molecules, the hydroxyl groups prefer the τ(hydroxyl) = 180° conformation in both compounds, with the OH bond being coplanar and syn with the ring C1-C6 bond. The constituent methyl groups prefer the staggered conformation, approximately at τ(methyl) = 60°.

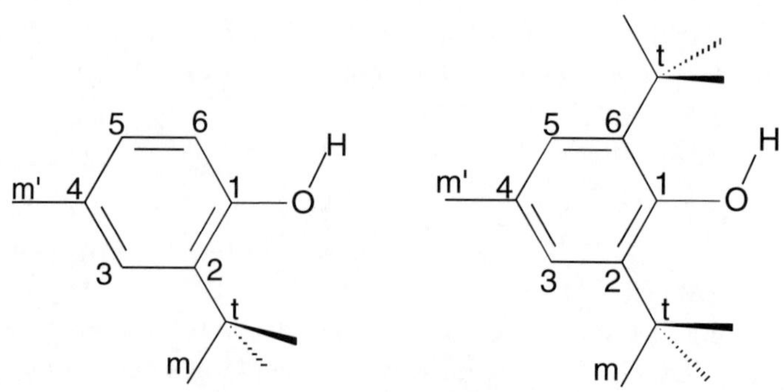

Scheme 4.1.

Table 4.1. Comparison between the calculated structure and the X-ray diffraction results of 2-*t*-butyl-4-methylphenol (1,2,4-OTMB)

Parameter	HF/6-31G(d)	X-ray	
		Site 1	Site 2
Bond length, Å			
C1C2	1.406	1.408	1.410
C2C3	1.387	1.391	1.393
C3C4	1.395	1.391	1.388
C4C5	1.378	1.385	1.390
C5C6	1.387	1.386	1.387
C6C1	1.380	1.379	1.382
C4Cm’	1.512	1.511	1.501
C1O	1.360	1.380	1.373
C2Ct	1.543	1.533	1.527
CtCm, in-plane[a]	1.538	1.532	1.529
CtCm, out-of-plane[b]	1.542(1.542)	1.539(1.534)	1.540(1.537)
Bond angle,°			
α(C2C1O)	119.2	118.4	118.6
α(C6C1O)	119.8	120.0	120.2
α(C3C2C1)	116.0	115.4	115.6
α(C3C2Ct)	122.0	121.9	122.3
α(C1C2Ct)	122.0	122.7	122.1
α(C2CtCm), in-plane[a]	111.9	111.5	112.1
α(C2CtCm), out-of-plane[b]	110.2(110.2)	111.0(109.3)	109.8(110.2)
Bond dihedral,°			
τ(C3C2CtCm), in-plane[a]	0.0	-9.2	-1.7
τ(C3C2CtCm), out-of-plane[b]	-119.2(119.2)	-128.4(110.0)	-120.3(118.1)

[a] In-plane or nearly in-plane methyl carbon.
[b] Two out-of-plane methyl carbons, one in parenthesis.

Table 4.2. Comparison between the calculated structure and the X-ray diffraction results of 2,6-di-*t*-butyl-4-methylphenol(1,2,4,6-OTMTB)

Parameter	HF/6-31G(d)	X-ray
Bond length, Å		
C1C2	1.399	1.402
C2C3	1.396	1.399
C3C4	1.378	1.382
C4C5	1.388	1.391
C5C6	1.385	1.395
C6C1	1.407	1.403
C4Cm′	1.512	1.515
C1O	1.362	1.390

Table 4.2. (Continued)

Parameter	HF/6-31G(d)	X-ray
C2Ct	1.551	1.539
CtCm, in-plane[a]	1.541	1.543
CtCm, out-of-plane[b]	1.547(1.547)	1.539(1.534)
C6Ct′	1.547	1.542
Ct′Cm, in-plane[a]	1.540	1.530
Ct′Cm, out-of-plane[b]	1.543(1.543)	1.541(1.534)
Bond angle,°		
C2CtCm, in-plane[a]	111.9	111.8
C2CtCm, out-of-plane[b]	111.1(111.1)	110.9(110.6)
C6Ct′Cm, in-plane[a]	111.7	112.2
C6Ct′Cm, out-of-plane[b]	110.6(110.6)	110.0(110.6)
Bond dihedral,°		
τ(C3C2CtCm), in-plane[a]	0.0	-5.3
τ(C3C2CtCm), out-of-plane[b]	-117.8(117.8)	-123.6(113.1)
τ(C5C6Ct′Cm), in-plane[a]	0.0	1.0
τ(C5C6Ct′Cm), out-of-plane[b]	-118.7(118.7)	-118.1(119.4)

[a] In-plane or nearly in-plane methyl carbon.
[b] Two out-of-plane methyl carbons, one in parenthesis.

The computed ground state structures of the isolated molecules were used as the starting point to accomplish the following studies on the internal rotation in the isolated molecules.

4.4.3. Internal Rotations of the *t*-Butyl Groups in the Isolated Molecules

The potential energy surfaces (PESs) of the internal rotation of the *t*-butyl groups in the isolated molecules of **1,2,4-OTMB** and **1,2,4,6-OTMTB** were calculated at the HF/6-311+G(d,p)//HF/6-31G(d) level. They are shown in Figure 4.1 in the range from τ(*t*-butyl) = 0 to 180°. There is a local three-fold symmetry with respect to the internal rotation of the *t*-butyl groups. The PESs in the range from τ(*t*-butyl) = 0 to 60° provide complete information about the internal rotation. The PESs in other regions can be obtained by the symmetry property. The calculations were carried out with the dihedral of τ(*t*-butyl) fixed at specific values and the rest of the molecules fully optimized at the HF/6-31G(d) level without symmetry constraint on the molecular geometry. The optimizations were followed by a single point energy calculation at the HF/6-311+G(d,p) level.

As seen in Figure 4.1, the ground state conformations of all the *t*-butyl groups are at τ(*t*-butyl) = 0°, in which one of the Cm-Ct bonds is coplanar and syn with the C2-C3 bond or the C6-C5 bond of the benzene ring. This preference reflects the fact that the *t*-butyl groups avoid close contact with the hydroxyl groups at the 1-position. By rotating the *t*-butyl groups around their Ct-Ca bonds by 60°, the rotational intermediates, *i.e.* the local energy minima, are obtained. The computed rotational transition states are approximately at τ(*t*-butyl) = 45°. But we notice that the energy differences between the transition states and the intermediate states are very small and there are flat tops in the PESs of all the three *t*-butyl groups. The barriers to the *t*-butyl rotations, defined as the energy difference between the transition state and the ground state, are 3.58 kcal/mol for the 2-*t*-butyl group in **1,2,4-OTMB**, and 4.08 and 3.09 kcal/mol for the 2-*t*-butyl and 6-*t*-butyl groups in **1,2,4,6-OTMTB**, respectively. The barriers are collected in Table 4.3 along with the barriers to the rotations of the constituent methyl groups. The potential energy surfaces can be fitted to the sum of a three-fold cosine potential and a six-fold cosine potential, $V(\tau)=\frac{V_3}{2}(1-\cos 3\tau)+\frac{V_6}{2}(1-\cos 6\tau)$. The least-square best-fit curves are plotted in Figure 4.1 in solid and dotted lines. The best-fit coefficients of V_3 and V_6 are collected in Table 4.4.

The constituent methyl groups of the *t*-butyl groups and the hydroxyl groups reorient during the internal rotation of the *t*-butyl groups. Along the internal rotation of the 2-*t*-butyl group in both **1,2,4-OTMB** and **1,2,4,6-OTMTB**, the largest structural changes occur at τ(2-*t*-butyl) = 30°, in which one of the Ct-Cm bonds is approximately perpendicular to the benzene ring. In **1,2,4-OTMB**, the hydroxyl group reorients by 2° to τ(hydroxyl) = 178° from 180° at τ(*t*-butyl) = 30° and then goes back to 180° in the rotational intermediate, τ(*t*-butyl) = 60°. The methyl group closer to the hydroxyl group changes its orientation by 9° from τ(methyl) = 60° to τ(methyl) = 51°. The methyl group further away from the hydroxyl group but close to the ring changes about 6° to τ(methyl) = 66°. The third methyl group, in which the Ct-Cm bond has a dihedral of τ(C3C2CtCm) = 90°, (called the perpendicular methyl group hereafter) only reorients about 1°. At the *t*-butyl rotational intermediate state, τ(*t*-butyl) = 60°, the methyl groups are back to the τ(methyl) dihedrals they had in the ground state.

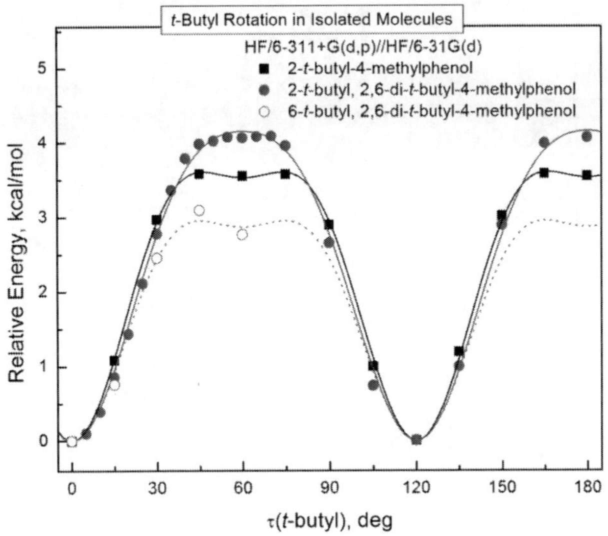

Figure 4.1. The fully relaxed rotational potential energy surfaces for the internal rotation of *t*-butyl groups in the isolated molecules, calculated at the HF/6-311+G(d,p)//HF/6-31G(d) level. The curves are the least-square best-fit results with the function, $V(\tau)=\frac{V_3}{2}(1-\cos 3\tau)+\frac{V_6}{2}(1-\cos 6\tau)$.

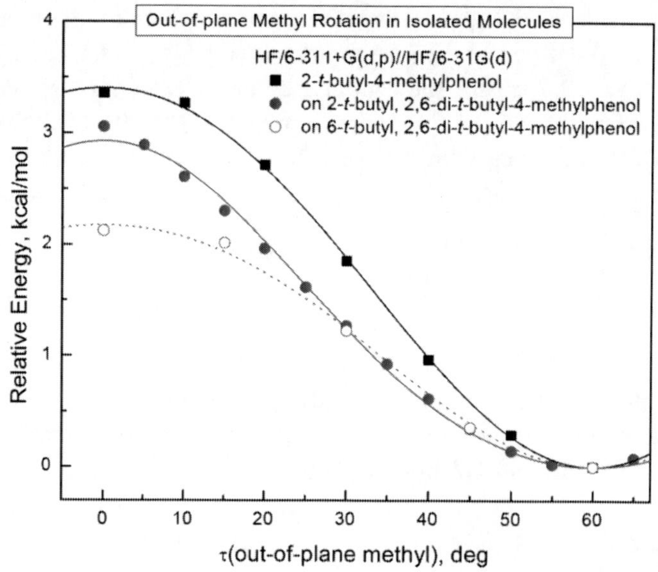

Figure 4.2. The fully relaxed rotational potential energy surfaces of the out-of-plane methyl groups in the isolated molecules, calculated at the HF/6-311+ G(d,p)//HF/6-31G(d) level. The curves are the least-square best-fit with the function, $V(\tau)=\frac{V_3}{2}(1-\cos 3(\tau-\frac{\pi}{3}))+\frac{V_6}{2}(1-\cos 6\tau)$.

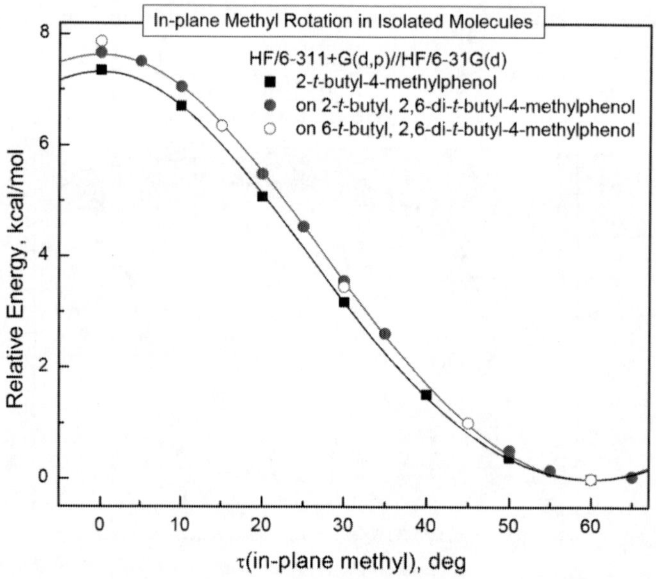

Figure 4.3. The fully relaxed rotational potential energy surfaces of the in-plane methyl groups in the isolated molecules, calculated at the HF/6-311+G(d,p)//HF/6-31G(d) level. The curves are the least-square best-fit with the function, $V(\tau)=\frac{V_3}{2}(1-\cos 3(\tau-\frac{\pi}{3}))+\frac{V_6}{2}(1-\cos 6\tau)$.

Table 4.3. Fully relaxed rotation barriers in the isolated molecules computed at the HF/6-311+G(d,p)//HF/6-31G(d) level

Rotor	Barrier, kcal/mol	
	2-*t*-butyl-4-methylphenol	**2,6-di-*t*-butyl-4-methylphenol**
Hydroxyl	3.11	0.46
t-butyl	3.58	4.08(3.09)[a]
Out-of-plane methyl	3.36	3.06(2.12)[a]
In-plane methyl	7.34	7.66(7.87)[a]

[a] Values in parenthesis refer to the 6-*t*-butyl group or the constituent methyl groups of the 6-*t*-butyl group (closer to the hydroxyl hydrogen).

Table 4.4. The least-square best-fit results of the three-fold (V_3) and six-fold (V_6) components for the rotational PESs of the *t*-butyl groups and the constituent methyl groups in the isolated molecules

Rotor	V_3, kcal/mol	V_6, kcal/mol
t-butyl[a]		
1,2,4-OTMB	3.53 ± 0.03	1.18 ± 0.03
1,2,4,6-OTMTB, 2-*t*-butyl[c]	4.15 ± 0.04	0.70 ± 0.04
1,2,4,6-OTMTB, 6-*t*-butyl[c]	2.87 ± 0.12	1.01 ± 0.14
Out-of-plane methyl[b]		
1,2,4-OTMB	3.40 ± 0.03	0.17 ± 0.03
1,2,4,6-OTMTB, 2-*t*-butyl[c]	2.93 ± 0.03	-0.22 ± 0.04
1,2,4,6-OTMTB, 6-*t*-butyl[c]	2.17 ± 0.05	0.16 ± 0.06
In-plane methyl[b]		
1,2,4-OTMB	7.31 ± 0.03	-0.48 ± 0.03
1,2,4,6-OTMTB, 2-*t*-butyl[c]	7.63 ± 0.02	-0.27 ± 0.03
1,2,4,6-OTMTB, 6-*t*-butyl[c]	7.80 ± 0.07	-0.45 ± 0.08

[a] Fitted with $V(\tau)=\frac{V_3}{2}(1-\cos 3\tau)+\frac{V_6}{2}(1-\cos 6\tau)$.

[b] Fitted with $V(\tau)=\frac{V_3}{2}(1-\cos 3(\tau-\frac{\pi}{3}))+\frac{V_6}{2}(1-\cos 6\tau)$.

[c] The 6-*t*-butyl group is the one closer to the hydroxyl hydrogen.

Similar changes occur during the internal rotation of the 2-*t*-butyl group in **1,2,4,6-OTMTB**. At τ(2-*t*-butyl) = 30°, the hydroxyl group reorients by 7° to τ(hydroxyl) = -173° from -180° in the ground state. The methyl group closer to the hydroxyl group adjusts its orientation to τ(methyl) = 73° from the staggered conformation in the ground state with τ(methyl) = 60°, and the methyl group further away from the hydroxyl group but close to the ring changes its orientation to τ(methyl) = 55° from 60°. The perpendicular methyl group retains its staggered conformation. Similar structural changes occur to the methyl groups along the rotation of the 6-*t*-butyl in **1,2,4,6-OTMTB**. However, due to the close contact between the hydroxyl group and the 6-*t*-butyl group, the orientation of the hydroxyl group changes dramatically from the ground state conformation τ(hydroxyl) = 180° to τ(hydroxyl) = 132° at τ(6-*t*-butyl) = 30°, to 151° at τ(6-*t*-butyl) = 45°, and then to t(hydroxyl) =149° at the intermediate state with τ(6-*t*-butyl) = 60°.

4.4.4. Internal Rotations of the Constituent Methyl Groups in the Isolated Molecules

The potential energy surfaces of the internal rotation of the constituent methyl groups of the *t*-butyl groups were also calculated at the HF/6-311+G(d,p)//HF/6-31G(d) level in the isolated molecules of the two compounds. In the ground state orientation of the *t*-butyl groups, there are two kinds of the methyl groups characterized by τ(*t*-butyl) = 0° and by τ(*t*-butyl) = ±120°. For the former type of methyl group, the Cm-Ct bond is coplanar with the ring and it is called the in-plane methyl group. The other two methyl groups are called the out-of-plane methyl groups. In Figure 4.2, the PESs of the out-of-plane methyl rotations are shown in the range from τ(methyl) = 0 to 60°. In Figure 4.3, the PESs of the in-plane methyl rotations are drawn in the range from τ(methyl) = 0 to 60°. The PESs at other regions can be obtained through the three-fold symmetry property of the methyl groups. For all methyl groups, the staggered conformation at τ(methyl) =60° is the ground state. After a rotation of 60° around the Cm-Ct bond, the transition state, which is the eclipsed conformation, is then reached and τ(methyl) = 0°.

The barrier to the out-of-plane methyl rotations in **1,2,4-OTMB** is 3.36 kcal/mol. The barriers in **1,2,4,6-OTMTB** are slightly lower, 3.03 kcal/mol for the out-of-plane methyl groups on the 2-*t*-butyl group and 3.25 kcal/mol on the 6-*t*-butyl group. The barrier to the in-plane methyl rotation in **1,2,4-OTMB** is 7.34 kcal/mol. The barriers to the in-plane methyl rotation on the 2-*t*-butyl and 6-*t*-butyl groups in **1,2,4,6-OTMTB** are slightly higher, 7.66 and 7.69 kcal/mol, respectively. The PESs can also be fitted with the sum of a three-fold cosine potential and a six-fold cosine potential, $V(\tau)=\frac{V_3}{2}(1-\cos 3(\tau-\frac{\pi}{3}))+\frac{V_6}{2}(1-\cos 6\tau)$. The least-square best-fit curves are plotted in Figures 4.2 and 4.3 in solid or dotted lines. The best-fit coefficients are collected in Table 4.4.

During the rotation of the out-of-plane methyl groups in **1,2,4-OTMB**, the hydroxyl group is not disturbed, staying at its ground state orientation, τ(hydroxyl) = 180°. The *t*-butyl group increases its dihedral slightly by 2° to τ(*t*-butyl) = -122° from τ(*t*-butyl) = -120°. However, for the out-of-plane methyl rotations on the 2-*t*-butyl group in **1,2,4,6-OTMTB**, not only is there a small magnitude of change of the *t*-butyl orientation from τ(*t*-butyl) = -120° to -123° at the transition state, but also the hydroxyl group reorients by 8° from τ(hydroxyl) = -180° to -172°. In the staggered conformation (τ(methyl) = 60°) of the out-of-plane methyl groups on the 6-*t*-butyl group in **1,2,4,6-OTMTB**, the optimized τ(*t*-butyl) is about 117°, which is 3° reorientation from the ground state orientation (τ(*t*-butyl) = 120°), and the hydroxyl group is at τ(hydroxyl) = -180°. After the methyl group rotates by 60° to the transition state (τ(methyl) = 0°), the *t*-butyl group reorients slightly to τ(*t*-butyl) = 122°. The hydroxyl group reorients by 18° to τ(hydroxyl) = -162° from τ(hydroxyl) = -180°.

During the internal rotation of the in-plane methyl group on the 2-*t*-butyl group in **1,2,4-OTMB**, the *t*-butyl group stays at its ground state orientation, τ(*t*-butyl) = 0°. The hydroxyl group also remains at the ground state orientation, τ(hydroxyl) = -180°. However, a small magnitude of change is found at τ(methyl) = 30°, in which the *t*-butyl group turns slightly to τ(*t*-butyl) = -8.1° and the hydroxyl reorients to τ(hydroxyl) = -178°. In addition, at the methyl rotation transition state, τ(methyl) = 0°, the bond angles of the *t*-butyl group, α(CmCtC2) and α(CtC2C3), increase slightly to 124° from 122° and to 114° from 112° in the ground state, respectively. Similar changes accompany the rotation of the in-plane methyl group on the 2-*t*-butyl group in **1,2,4,6-OTMTB**. The bond angles, α(CmCtC2) and α(CtC2C3), increase by 2° at the methyl rotational transition state, τ(methyl) = 0°. At τ(methyl) = 30°, the *t*-butyl group reorients by 9° to τ(*t*-butyl) = 9° from 0° and the hydroxyl group reorients by 10° to τ(hydroxyl) = -170° from -180°. They return to the ground state orientation in the methyl rotational transition state (τ(methyl) = 0°). Similar changes, but in a much larger magnitude, occur along the in-plane methyl rotation in the 6-*t*-butyl group. At τ(methyl) = 30°, the *t*-butyl group adjusts by 14° to τ(*t*-butyl) = 14° from 0° and the hydroxyl group responds by reorienting 32° to τ(hydroxyl) = 148° from 180°. At the transition state, τ(methyl) = 0°, the *t*-butyl group and the hydroxyl group return to the ground state orientations, but the bond angles α(CmCtC2) and α(CtC2C3), increase by 2°.

4.4.5. Internal Rotations of the Hydroxyl Group in the Isolated Molecules

The PESs of the internal rotation of the hydroxyl groups were also calculated at the HF/6-311+G(d,p)//HF/6-31G(d) level in the isolated molecules. They are shown in Figure 4.4 in the range from τ(hydroxyl) = 0 to 180°. The values in other regions can be obtained from symmetry. The hydroxyl rotation in **1,2,4-OTMB** is symmetric about the benzene ring plane. In **1,2,4,6-OTMTB**, the two *t*-butyl groups form a symmetric environment for the hydroxyl group rotation. There is an additional two-fold rotation symmetry. As shown earlier, in both the compounds, the conformation at τ(hydroxyl) = 180° is the ground state. For **1,2,4,6-OTMTB**, the conformation at τ(hydroxyl) = 0° is equivalent to 180° conformation, *i.e.* the ground state. In the τ(hydroxyl) = 0° conformation of **1,2,4-OTMB**, the hydroxyl hydrogen gets closer to the 2-*t*-butyl group and is located in between the two out-of-plane methyl groups. This is a rotational intermediate state for the hydroxyl group. The transition state occurs at τ(hydroxyl) = 60°. The transition state in **1,2,4,6-OTMTB** is also at τ(hydroxyl) = 60°. The τ(hydroxyl) = 90° conformation of **1,2,4,6-OTMTB** is an intermediate state. The rotational barrier in **1,2,4-OTMB** is 3.11 kcal/mol, but is only 0.46 kcal/mol in **1,2,4,6-OTMTB**. The PES of the hydroxyl rotation in **1,2,4-OTMB** can be fitted into the sum of a one-fold cosine potential and a two-fold cosine potential, $V(\tau)=\frac{V_1}{2}(1-\cos(\tau-\pi))+\frac{V_2}{2}(1-\cos 2\tau)$, while in **1,2,4,6-OTMTB**, only the two-fold cosine potential is needed, $V(\tau)=\frac{V_2}{2}(1-\cos 2\tau)$. However, to account the details of the

PESs, the four-fold cosine potential is added in, $V(\tau)=\frac{V_1}{2}(1-\cos(\tau-\pi))+\frac{V_2}{2}(1-\cos 2\tau)+\frac{V_4}{2}(1-\cos 4\tau)$. The least-square best-fit curves are plotted in Figure 4.4 in solid lines. The best-fit coefficients are collected in Table 4.5.

Table 4.5. The least-square best fit coefficients of the hydroxyl rotational potential energy surfaces in the isolated molecules with the function,

$$V(\tau)=\frac{V_1}{2}(1-\cos(\tau-\pi))+\frac{V_2}{2}(1-\cos 2\tau)+\frac{V_4}{2}(1-\cos 4\tau)$$

Compound	V_1, kcal/mol	V_2, kcal/mol	V_4, kcal/mol
1,2,4-OTMB	2.72 ± 0.02	1.37 ± 0.02	0.10 ± 0.02
1,2,4,6-OTMTB	0.00 ± 0.003	0.44 ± 0.004	0.16 ± 0.004

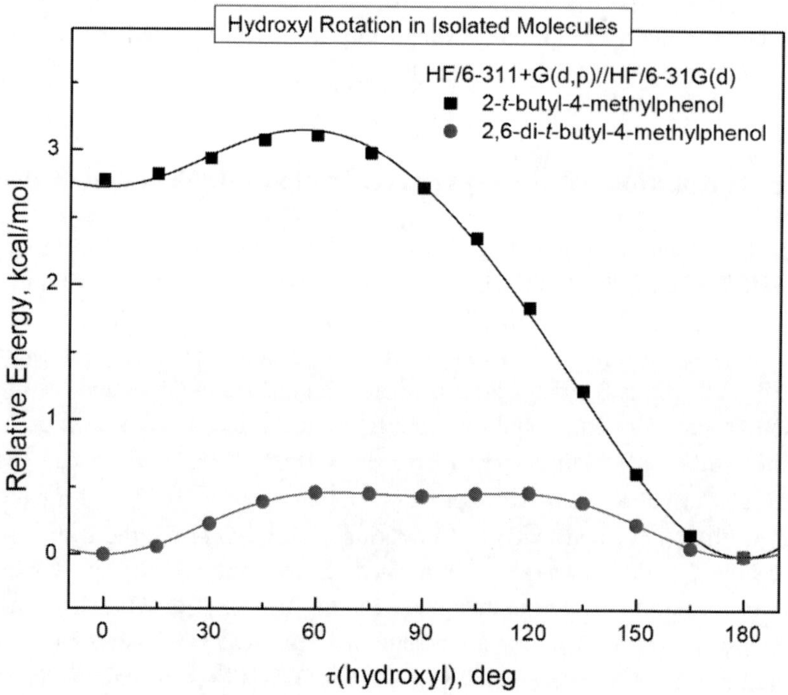

Figure 4.4. The fully relaxed rotational potential energy surfaces of the hydroxyl groups in the isolated molecules, calculated at the HF/6-311+G(d,p)//HF/6-31G(d) level. The curves are the least-square best fit to, $V(\tau)=\frac{V_1}{2}(1-\cos(\tau-\pi))+\frac{V_2}{2}(1-\cos 2\tau)+\frac{V_4}{2}(1-\cos 4\tau)$. The best-fit coefficients are collected in Table 4.5.

Small structural relaxations are found with the hydroxyl rotation. The first relaxation is that the hydroxyl group opens its α(HOC1) and α(OC1C6) bond angles slightly. The bond angles of α(HOC1) and α(OC1C6) are 110 and 120° in the ground state in **1,2,4-OTMB**. In its intermediate sate at τ(hydroxyl) = 0°, the bond angles of α(HOC1) and α(OC1C2) increase slightly to 112 and 125°, respectively. At the hydroxyl rotational transition state, the 2-*t*-butyl group reorients by 3° to τ(*t*-butyl) = -3° from 0°. The largest relaxation of the methyl groups is observed at τ(hydroxyl) = 60°. The out-of-plane methyl group closer to the hydroxyl hydrogen reorients by 8° to τ(methyl) = 68° from the staggered conformation (τ(methyl) = 0°). In **1,2,4,6-OTMTB**, larger magnitude of the structural relaxation of the methyl groups is found along the hydroxyl group rotation. At τ(hydroxyl) = 90°, the two out-of-plane methyl groups on the same side with the hydroxyl hydrogen reorient by 9° to τ(methyl) = 69 and -69° from 60 and -60°, respectively. At τ(hydroxyl) = 60°, the relaxation reaches to its maximum. The methyl group closer to the hydroxyl hydrogen reorients by 14° to τ(methyl) = 74° from 60°. The other reoriented methyl group gets back slightly to τ(methyl) = -64°. Other structural changes are negligible except the C1-O bond length. In both the compounds, if the hydroxyl group is coplanar with the ring, τ(hydroxyl) = 0° or 180°, the C1-O bond is about 1.360 Å and when the hydroxyl group forms a right angle with the ring, τ(hydroxyl) = ±90°, the bond length elongates to about 1.380 Å.

4.4.6. Two-dimensional Potential Energy Surfaces of the Internal Rotations of the *t*-Butyl Group and Its Constituent Methyl Groups in the Isolated Molecules

To further explore the rotational coupling relationship between the *t*-butyl group and the constituent methyl groups, the two-dimensional PESs were calculated with respect to the rotational coordinates of the *t*-butyl group and the related methyl group sharing the same Cm for τ(*t*-butyl) and τ(methyl). The PESs, calculated at the HF/6-311+G(d,p)//HF/6-31G(d) level, are shown in Figure 4.5 for the 2-*t*-butyl group in **1,2,4-OTMB** and in Figure 4.6 for the 2-*t*-butyl group in **1,2,4,6-OTMTB**. (The calculation for the 6-*t*-butyl group in **1,2,4,6-OTMTB** is not useful since the strong intramolecular interactions will drive the rotation of the hydroxyl group. The calculation with the fixed orientation of the hydroxyl group is given in a later section.) In the PESs, only the range from τ(methyl) = 0 to 60° and τ(*t*-butyl) = 0 to 180° is shown. Other regions can be obtained by symmetry. The PES of **1,2,4-OTMB** was calculated for every 7.5° along the coordinates τ(methyl) and τ(*t*-butyl). The PES of **1,2,4,6-OTMTB** was calculated to a lower resolution, every 15° along the coordinate of τ(methyl) and every 30° along the coordinate of τ(*t*-butyl).

As seen from Figures 4.5 and 4.6, the two PESs are similar. We focus our discussion on the PES of **1,2,4-OTMB**. The conclusions can then be extended to the isolated molecule of **1,2,4,6-OTMTB**. Figure 4.5 shows that the barrier to *t*-butyl rotation increases to 3.5 kcal/mol due to the presence of the ortho hydroxyl group. This high barrier to *t*-butyl rotation isolates the two troughs of the in-plane methyl (τ(*t*-butyl) = 0°) and out-of-plane methyl (τ(*t*-butyl) = 120°) rotational paths. The out-of-plane methyl rotation has a similar barrier height as does the *t*-butyl group, while the barrier to the in-plane methyl rotation is much higher at about 7.5 kcal/mol. At room temperature, it is likely that the rotations of the out-of-plane methyl groups and the *t*-butyl group occur with a much higher probability than the in-plane methyl rotation. Also in Figure 4.5, we see it is unlikely that the methyl rotations are coupled with the *t*-butyl rotation, but the small magnitudes of orientation relaxation are not excluded since the neighboring regions along the minimum energy paths are quite flat.

The PESs can also be viewed as the dependence of the "constrained" methyl rotation on the *t*-butyl orientation. (**) We call the methyl rotation along the straight path parallel to the τ(methyl) axis as the "constrained" rotation. Each constrained methyl rotation follows roughly a three-fold potential energy curve, but the barrier is dependent on the *t*-butyl orientation, τ(*t*-butyl). In Figure 4.7, the dependence is plotted for the two molecules. The curves for the two PESs follow a same trend, although the constrained in-plane rotation barrier in **1,2,4,6-OTMTB** is much higher than that in **1,2,4-OTMB**. The lowest methyl rotation barrier is achieved at τ(*t*-butyl) =135°, slightly off the out-of-plane methyl position in the ground state, τ(*t*-butyl) =120°. The dependence curves can be fitted into the sum of a few cosine functions, $V_m(\tau_t) = V_{m0}1 + V_{m1}\cos(\tau_t - \pi) + V_{m2}\cos 2\tau_t + V_{m3}\cos 3(\tau_t - \pi) + V_{m4}\cos 4\tau_t$. The least-square best-fit results are plotted in Figure 4.7 in solid lines. The best-fit coefficients are collected in Table 4.6.

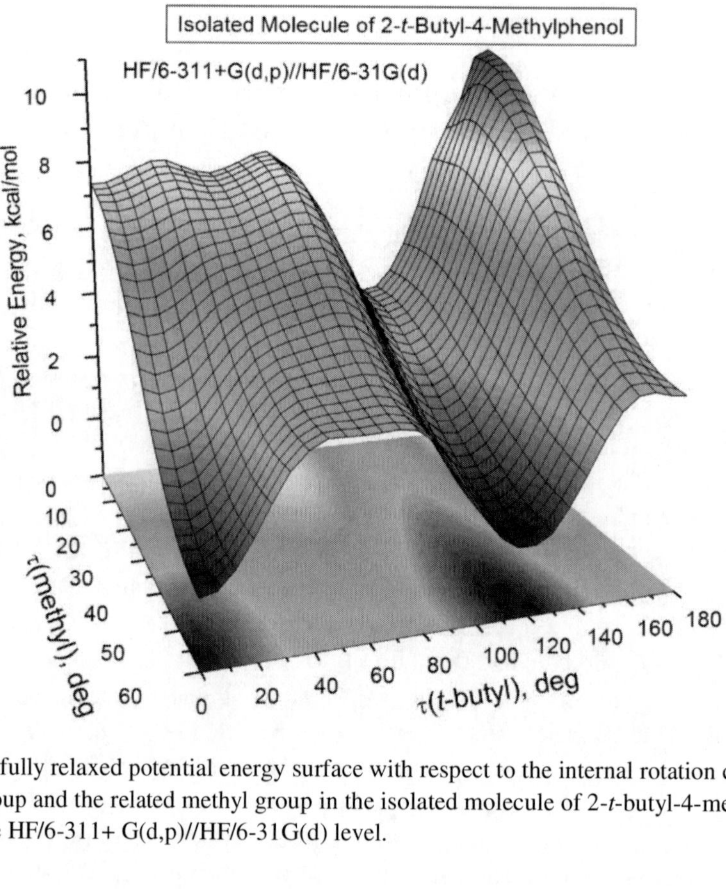

Figure 4.5. The fully relaxed potential energy surface with respect to the internal rotation coordinates of the 2-*t*-butyl group and the related methyl group in the isolated molecule of 2-*t*-butyl-4-methylphenol, calculated at the HF/6-311+ G(d,p)//HF/6-31G(d) level.

** See the previous two chapters for more discussion on this analysis.

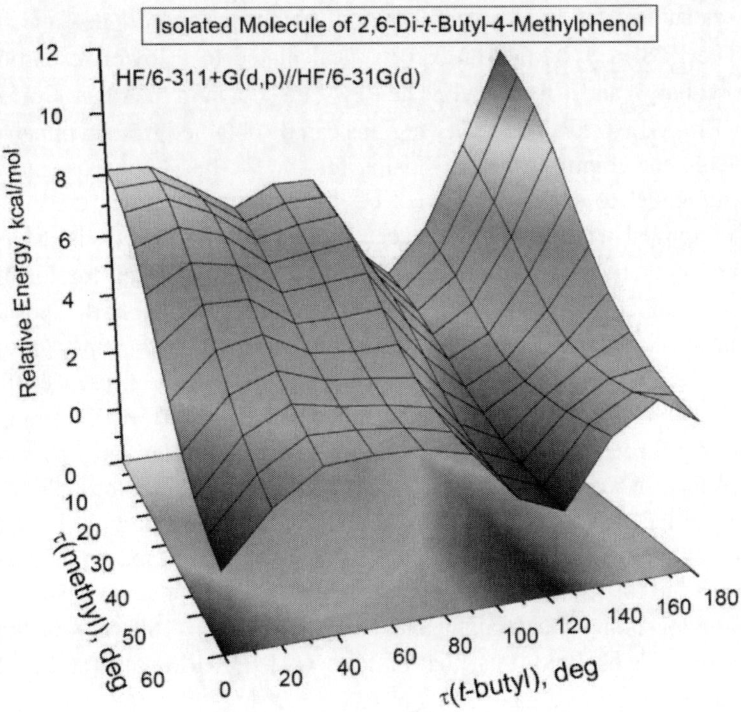

Figure 4.6. The fully relaxed potential energy surface with respect to the internal rotation coordinates of the 2-*t*-butyl group and the related methyl group in the isolated molecule of 2,6-di-*t*-butyl-4-methylphenol, calculated at the HF/6-311+ G(d,p)//HF/6-31G(d) level.

Table 4.6. The least square best fit coefficients of the dependence of the constrained methyl rotational barrier on the orientation of the 2-*t*-butyl group with the function,

$$V_m(\tau_t) = V_{m0}1 + V_{m1}\cos(\tau_t - \pi) + V_{m2}\cos 2\tau_t + V_{m3}\cos 3(\tau_t - \pi) + V_{m4}\cos 4\tau_t$$

Compound	V_{m0}	V_{m1}	V_{m2}	V_{m3}	V_{m4}
1,2,4-OTMB	9.50 ± 0.10	-1.03 ± 0.15	2.46 ± 0.15	0.40 ± 0.15	2.42 ± 0.15

4.4.7. Two-dimensional Potential Energy Surfaces with respect to Internal Rotations of the *t*-Butyl and Hydroxyl Groups in the Isolated Molecules

We have calculated the two-dimensional PESs of the two molecules with respect to the internal coordinates of the 2-*t*-butyl group and the hydroxyl group to investigate the coupling relationship between the two groups. The PESs, computed at the HF/6-311+G(d,p)//HF/6-31G(d) level, are shown in Figures 4.8 and 4.9 for the compounds of **1,2,4-OTMB** and **1,2,4,6-OTMTB**, respectively. (The information on the coupling relationship between the hydroxyl group and both the 2-*t*-butyl and 6-*t*-butyl groups is contained in the PES of **1,2,4,6-OTMTB**. The rotation of the hydroxyl group the *t*-butyl groups converts the spatial relationship between the two *t*-butyl groups and the hydroxyl group.) The PES of **1,2,4-**

OTMB was calculated at 7.5° interval along both the coordinates of τ(*t*-butyl) and τ(hydroxyl). The PES of **1,2,4,6-OTMTB** was calculated to a lower resolution, every 15° interval along τ(*t*-butyl) and τ(hydroxyl). The PESs are shown in the range from τ(*t*-butyl) = 0 to 60° and τ(hydroxyl) = 0 to 180°. Other regions can be obtained from symmetry.

In Figure 4.8, the minimum energy paths for the 2-*t*-butyl rotation and the hydroxyl rotation are orthogonal to each other. Each of them roughly follows their coordinate axis directions. The coupled rotation route between the two groups, which is diagonal from the global minimum at τ(*t*-butyl) = 0° and τ(hydroxyl) = 180° bypassing the local minimum at τ(*t*-butyl) = 60° and τ(hydroxyl) = 0°, is not favorable, although the possibility is not excluded for the small magnitude reorientation of one group when the other group reorients.

The PES of **1,2,4,6-OTMTB** changes dramatically due to the presence of the 6-*t*-butyl group. Both corners at τ(hydroxyl) = 0° and 180° when τ(*t*-butyl) = 0° are the ground state. There are two *t*-butyl rotation paths, one corresponding to the rotation of the 2-*t*-butyl group and the other to the 6-*t*-butyl group. Due to the presence of the steric-crowded *t*-butyl groups on both sides of the 1-hydroxyl group, the barrier to the hydroxyl rotation decreases significantly to about 0.5 kcal/mol. In return, this low barrier makes it possible for the coupling between the rotations of the 2-*t*-butyl group and the hydroxyl group. The barrier to the 2-*t*-butyl rotation along the straight path at τ(hydroxyl) = 180° is 4.06 kcal/mol. This value is higher by 0.48 kcal/mol than that in **1,2,4-OTMB**. However, the barrier to the 6-*t*-butyl rotation along the straight path at τ(hydroxyl) = 0° is only 2.76 kcal/mol. Although the barrier is also higher than the barrier to the 2-*t*-butyl rotation along the straight path at τ(hydroxyl) = 0° in **1,2,4-OTMB**, the path of the 6-*t*-butyl rotation starts at the ground state of **1,2,4,6-OTMTB** and it is indeed a minimum energy path for the *t*-butyl rotation. In addition, due to the low barrier to the rotation of the hydroxyl group, there is no strong hindrance to the 2-*t*-butyl group coupling with the rotation of the hydroxyl group and to follow a diagonal path. The coupling will lower the barrier significantly by about 1.2 kcal/mol. In the previous section, we have seen the minimum energy path for the 6-*t*-butyl group rotation does not follow the straight path at τ(hydroxyl) = 0° because the steric repulsion between the 6-*t*-butyl group and the hydroxyl group. Instead, the hydroxyl group reorients as large as 45° along the 6-*t*-butyl rotation.

The PESs can also be viewed as the dependence of the constrained *t*-butyl rotation on the hydroxyl group orientation. Here, the 'constrained' rotation refers to the fixed orientation of the hydroxyl group. In Figure 4.10, the barriers to the constrained *t*-butyl rotations are plotted against the hydroxyl orientation, τ(hydroxyl). Although the two PESs are very different, the dependence curves of the barriers follow a similar shape. The barriers in **1,2,4,6-OTMTB** are higher than those in **1,2,4-OTMB**. In the range from τ(hydroxyl) = 0 to 60°, the constrained *t*-butyl rotation barriers are almost constant. When τ(hydroxyl) increases from 60 to 120°, the barriers also increase proportionally. In the range from τ(hydroxyl) = 120 to 180°, the increasing trend continues, but at a smaller rate. The dependence curves can be fitted with the sum of a few cosine terms, $V_t(\tau_{OH}) = V_{t0}1 + V_{t1}\cos\tau_{OH}$. The least-square best-fit results are plotted in Figure 4.10 in solid lines. The best-fit coefficients are collected in Table 4.7. The contributions from the higher-order cosine components are negligible.

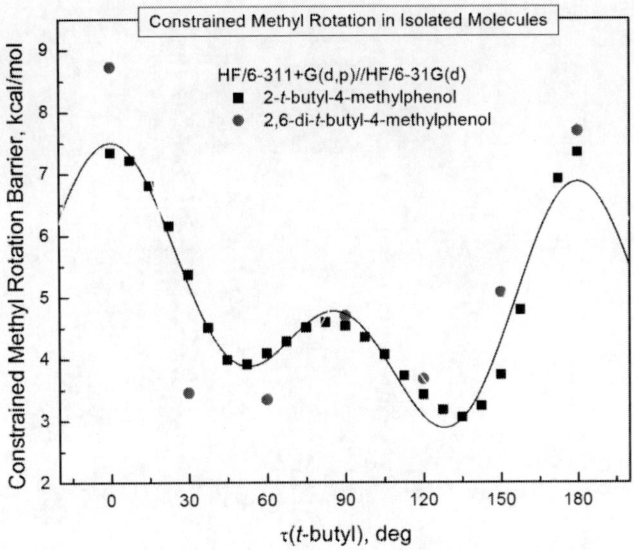

Figure 4.7. The dependence of the constrained rotation barrier of the methyl group on the relative orientation of the 2-*t*-butyl group, τ(*t*-butyl). The barriers are from the two-dimensional potential energy surfaces in Figures 4.5 and 4.6. The solid curve is the least-square best-fit to the function, $V_m(\tau_t) = V_{m0}1 + V_{m1}\cos(\tau_t - \pi) + V_{m2}\cos 2\tau_t + V_{m3}\cos 3(\tau_t - \pi) + V_{m4}\cos 4\tau_t$. The best-fit coefficients are collected in Table 4.6.

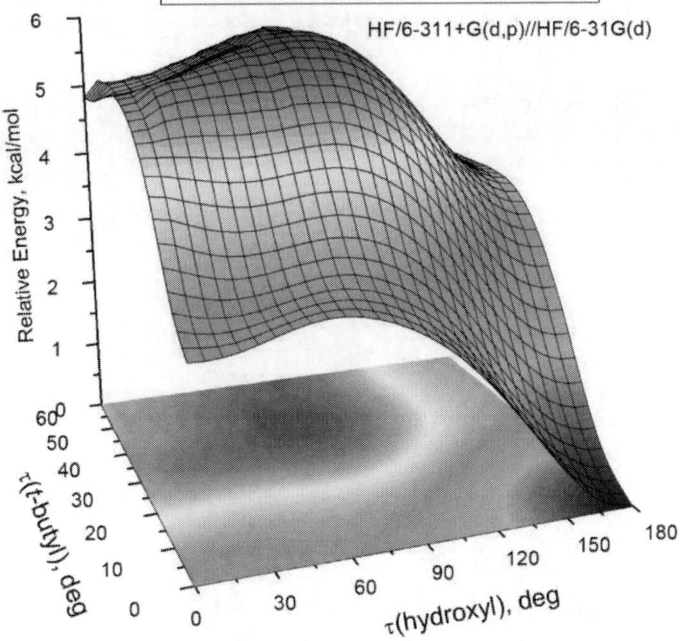

Figure 4.8. The fully relaxed potential energy surface with respect to the internal rotation coordinates of the 2-*t*-butyl group and the hydroxyl group in the isolated molecule of 2-*t*-butyl-4-methylphenol, calculated at the HF/6-311+G(d,p)//HF/6-31G(d) level.

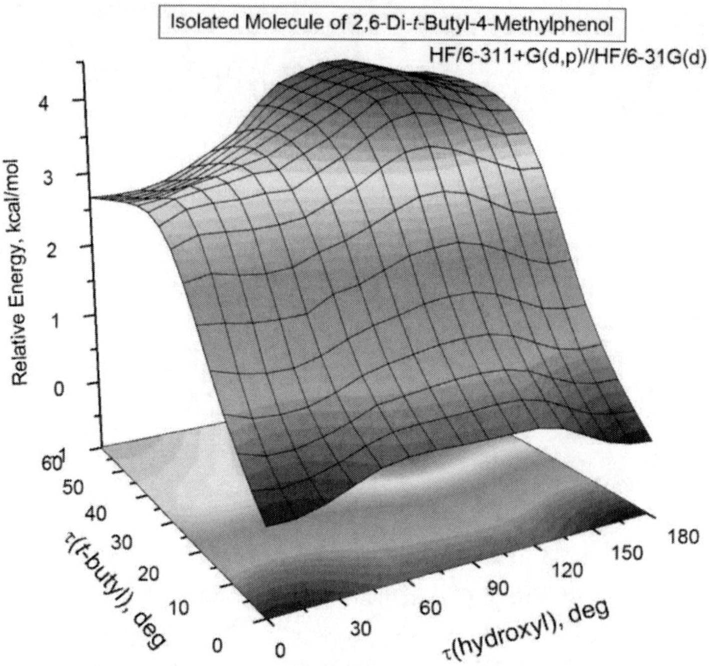

Figure 4.9. The fully relaxed potential energy surface with respect to the internal rotation coordinates of the 2-*t*-butyl group and the hydroxyl group in the isolated molecule of 2,6-di-*t*-butyl-4-methylphenol, calculated at the HF/6-311+G(d,p)// HF/6-31G(d) level.

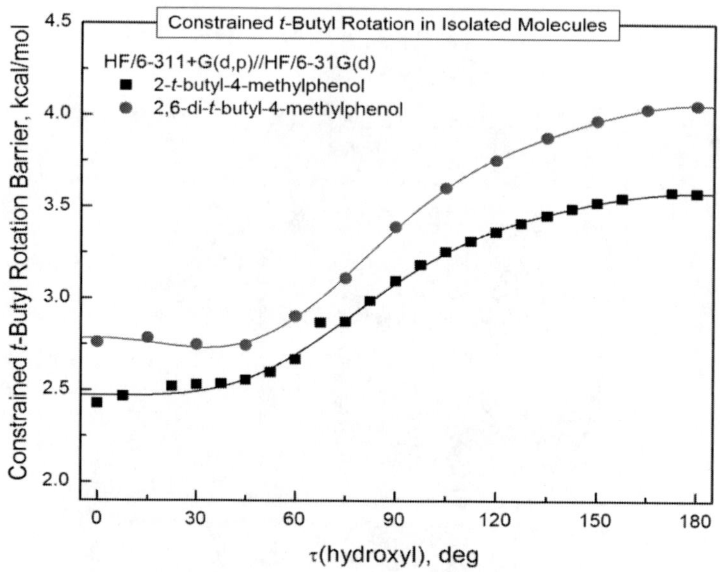

Figure 4.10. The dependence of the constrained rotation barrier of the *t*-butyl group on the relative orientation of the hydroxyl group, τ(hydroxyl) (= τ_{OH}). The barriers are from the two-dimensional potential energy surfaces in Figures 4.8 and 4.9. The solid curves are the least-square best fit with the function, $V_t(\tau_{OH}) = V_{t0}1 + V_{t1}\cos\tau_{OH}$.

Table 4.7. The least square best fit coefficients of the dependence of the constrained *t*-butyl rotational barrier on the orientation of the hydroxyl group with the function,

$$V_t(\tau_{OH}) = V_{t0}1 + V_{t1}\cos\tau_{OH}$$

Compound	V_{t0}	V_{t1}
1,2,4-OTMB	3.03 ± 0.01	0.59 ± 0.01
1,2,4,6-OTMTB	3.36 ± 0.02	0.71 ± 0.03

4.5. DISCUSSION

4.5.1. Effect of the Hydroxyl Group to the Internal Rotation of the *t*-Butyl Groups

The barrier to the internal rotation of the *t*-butyl groups in 1,4-di-*t*-butyl-benzene, 2,6-di-*t*-butylnaphthalene and 3-*t*-butylchrysene (they will be referred as the aromatic hydrocarbons) ranges from 0.5 to 1.0 kcal/mol in the isolated molecules. As discussed in Chapter 2, there is no strong steric hindrance to the internal rotation in the three aromatic hydrocarbons. The PES can be decomposed into the sum of a three-fold cosine component and a six-fold cosine component. The six-fold component is presumably due to the steric repulsion between the *t*-butyl group and the hydrogen atoms on the two ring carbons adjacent to the ring carbon (Ca) to which the *t*-butyl group is attached. The three-fold component is from the hyperconjugation between the *t*-butyl group and the aromatic ring. The six-fold component is about 0.5 kcal/mol, and roughly the same in the three aromatic systems. The three-fold component depends on the molecular structure, ranging from 0 to 0.5 kcal/mol. Furthermore, an empirical linear relationship is obtained between the three-fold component and the bond length difference of the two aromatic ring Ca-C bonds adjacent to the *t*-butyl group.

In **1,2,4-OTMB** and **1,2,4,6-OTMTB**, the barriers to the internal rotation of the *t*-butyl groups are much higher, in the range from 3.0 to 4.0 kcal/mol, due to the presence of the ortho hydroxyl group. The PES can also be decomposed into a three-fold cosine component and a six-fold cosine component. The six-fold component is similar in the two phenols, about 1.0 kcal/mol. The three-fold component, ranging from 2.9 to 4.1 kcal/mol, is three to four times larger than that in the three aromatic hydrocarbons. (See Table 4.4) The two ring carbons that are bonded to Ca are differently substituted. On one side, there is a hydroxyl group bonded to the ring C atom, and on the other side only a ring hydrogen. The six-fold component function has its maximum at 30° (the perpendicular conformation for the *t*-butyl group), and the origin of this component can be explained by the interaction between the constituent methyl groups and the ring hydrogen on one side and the hydroxyl group on the other side. (Most of the latter interaction contributes to the three-fold component.) This interaction is slightly stronger in the phenols than in the aromatic hydrocarbons, indicated by the larger dihedral changes of the methyl groups at τ(*t*-butyl) = 30°. The stronger six-fold component in the phenols can be explained by that the *t*-butyl groups have to open their bond angles (α(CtCaC1)) slightly on the hydroxyl group side and this causes stronger repulsion on the ring hydrogen side. The three-fold component reflects significant hindrance caused by the hydroxyl group to the *t*-butyl group rotation.

The presence of the hydroxyl group alters the *t*-butyl group rotational potential energy surfaces in two ways. One way is that the hydroxyl group brings steric hindrance to the *t*-butyl rotation. The other way is via the hyperconjugation between the oxygen lone electron pairs and the aromatic π orbitals [9-10, 27-29]. We discuss the latter effect in terms of the natural bond orbital (NBO) analysis [30-37].

There are two lone-electron-pair orbitals on the oxygen atom of the hydroxyl group. One is in-plane with the ring with the $sp^{1.4}$ character (lp(1)) and the other is orthogonal to the ring plane with the p_z character (lp(2)). See Scheme 4.2. The electron density on the lone pair orbital lp(2) is able to delocalize onto the antibonding π* orbitals of the benzene ring. The hyperconjugation causes different bond lengths of the benzene ring. For example, in **1,2,4-OTMB**, the calculated C1-C2 bond is longer than the C2-C3 bond by 0.018 Å. In **1,2,4,6-OTMTB**, the C1-C2 bond is longer than the C2-C3 bond by 0.022 Å. At the same time, the difference between the two bonds adjacent to the 6-*t*-butyl group (the C6-C1 bond and the C6-C5 bond) is negligible.

The NBO analysis shows that the lone pair orbital lp(2) with the p_z character shows a strong hyperconjugation (the second order perturbation interaction energy, $E^{(2)}$, is about 30~32 kcal/mol) with the antibonding π* orbital localized on the C1-C6 bond in both the phenols. There is a weaker but still significant hyperconjugation ($E^{(2)}$ is about 7 kcal/mol) between the other lone pair orbital lp(1) with the $sp^{1.4}$ character and the antibonding σ* orbital of the C1-C6 bond. The sum of these two hyperconjugation factors exceeds the repulsive interactions between the *t*-butyl groups and the hydroxyl group in **1,2,4,6-OTMTB** and determines the ground state is the in-plane conformation (τ(hydroxyl) = 0 or 180°) instead of the perpendicular conformation (τ(hydroxyl) = 90°). (The hyperconjugation favors the in-plane conformation and the steric effect from the *t*-butyl groups favors the perpendicular conformation. In the unsubstituted phenol molecule (**OB**), there is no steric effect from the *t*-butyl groups and the ground state is the in-plane conformation. However, the hydroxyl rotation barrier is 2.5 kcal/mol in **OB**, which is 2.0 kcal/mol higher than in **1,2,4,6-OTMTB**. The much lower barrier in **1,2,4,6-OTMTB** is because the hyperconjugation effect is offset by the steric effect of the *t*-butyl groups.)

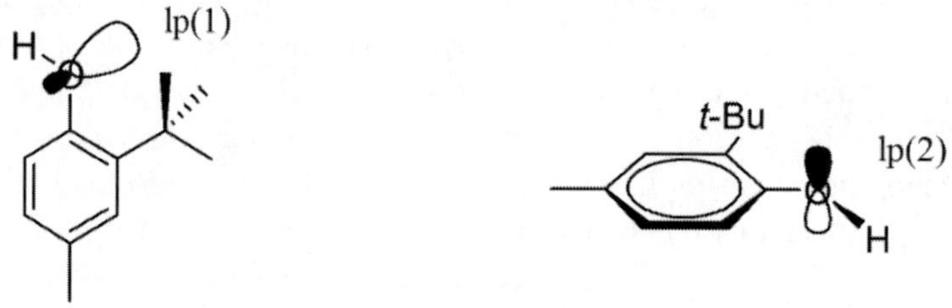

Scheme 4.2.

In **1,2,4-OTMB**, the repulsive interaction between the 2-*t*-butyl group and the lone pair lp(1) is stronger than the repulsion between the 2-*t*-butyl group and the hydroxyl hydrogen atom, and this off-balance causes the τ(hydroxyl) = 180° conformation is the ground state and the τ(hydroxyl) = 0° conformation is an intermediate state for the hydroxyl group rotation.

Since both the hyperconjugation and the steric repulsion from the hydroxyl groups contribute to the barriers in the three-fold component of the *t*-butyl rotation PES, it is impossible to distinguish them in the potential energy surfaces of the *t*-butyl rotations. However, we can make a rough estimate from the dependence curve of the constrained rotation barrier of the *t*-butyl group on the orientation of the hydroxyl group in Figure 4.10. As discussed earlier, the curve can be fitted to, $V_t(\tau) = V_{t0} + V_{t1}\cos\tau = (V_{t0} - V_{t1}) + (2V_{t1})\frac{1}{2}(1-\cos\tau)$. The first collected term, $(V_{t0} - V_{t1})$, can be viewed as the basic *t*-butyl barrier at τ(hydroxyl) = 0°, which is mainly due to the repulsion between the *t*-butyl group and the hydroxyl hydrogen. The second term is 0 at τ(hydroxyl) = 0° and reaches its maximum, $2V_{t1}$, at τ(hydroxyl) = 180°. The hyperconjugation effect of the hydroxyl group is negligible compared with the *t*-butyl rotation barrier at τ(hydroxyl) = 0° (*i.e.* 6-*t*-butyl group in **1,2,4,6-OTMTB**), which is evidenced by the equal bond length of the two ring bonds adjacent to the *t*-butyl groups. At τ(hydroxyl) = 180° (*i.e.* 2-*t*-butyl group), the hyperconjugation effect is maximum. If we use the linear relationship obtained in Chapter 2 for the aromatic hydrocarbons between the hyperconjugation contribution to the *t*-butyl rotation barriers and the bond length difference, 18 kcal/mol/Å, the hyperconjugation contribution is about 0.4 kcal/mol for the 2-*t*-butyl group rotation in **1,2,4-OTMB** and 0.3 kcal/mol for the 2-*t*-butyl group rotation in **1,2,4,6-OTMTB**. As shown in Table 4.7, the $2V_{t1}$ term is about 1.2 kcal/mol in **1,2,4-OTMB** and 1.4 kcal/mol in **1,2,4,6-OTMTB**. The stronger repulsion from the lp(1) of the oxygen contributes 0.8 kcal/mol (=1.2 - 0.4) in the *t*-butyl rotation barrier in **1,2,4-OTMB** and 1.1 kcal/mol (=1.4 - 0.3) in **1,2,4,6-OTMTB**.

4.5.2. Effect of the Hydroxyl Group on the Internal Rotation of the Methyl Groups in the *t*-Butyl Groups

The internal rotation of the methyl groups in the *t*-butyl groups is affected in two ways by the ortho hydroxyl groups. First, the in-plane methyl rotation can not be converted to an out-of-plane methyl rotation as in the isolated molecules of the aromatic hydrocarbons due to the high barrier to the *t*-butyl rotation caused by the hydroxyl hindrance. Second, the internal rotation of the out-of-plane methyl groups is also directly affected by the interactions from the hydroxyl group. As seen in Table 4.3, the barriers to the out-of-plane methyl rotation are smaller than those in the aromatic hydrocarbons in Chapter 2. Also because of the hydroxyl group, the *t*-butyl group relaxation which opens the bond angles α(CmCtC2) and α(CtC2C1) (or α(CmCtC6) and α(CtC6C5)) is limited for the in-plane methyl rotation. The barriers increase slightly compared with those in the three aromatic hydrocarbons.

The potential energy surfaces of the methyl rotation can be decomposed to a three-fold cosine potential and a six-fold cosine potential. See Table 4.4. The six-fold coefficients are small. They reflect the effect of the structural relaxation of the *t*-butyl group and the hydroxyl group in the methyl rotation.

The two-dimensional PESs with respect to the internal rotations of the *t*-butyl group and the related methyl group are similar for the two phenols. The dependence curve of the constrained rotation barrier for the methyl group on the orientation of the *t*-butyl group is similar with the relationship curve found in the aromatic hydrocarbons, although some details are different as shown in Figure 4.7. Because of the presence of the hydroxyl group, the curve is not symmetric about τ(*t*-butyl) = 90°. The minimum barrier does not appear at the out-of-plane methyl position near τ(*t*-butyl) = 120° but at τ(*t*-butyl) = 135°, and the barrier decreases by about 1.0 kcal/mol from the characteristic barrier of 4.0 kcal/mol in the aromatic hydrocarbons. But the in-plane methyl rotation barrier increases by 0.5 to 1.0 kcal/mol.

4.5.3. Effect of the *t*-Butyl Groups on the Internal Rotation of the Hydroxyl Groups

The hydroxyl rotation in phenol (**OB**) has been studied previously [9]. The conformation with the hydroxyl group coplanar with the benzene ring is the ground state and the perpendicular conformation is the transition state. The barrier is 2.5 kcal/mol computed at the MP2/6-31G(d) level [9], consistent with the experimental results [11-15]. In **1,2,4-OTMB**, the introduction of the ortho *t*-butyl group breaks the two-fold symmetry for the hydroxyl rotation in phenol. Not only is the transition state shifted to τ(hydroxyl) = 60°, but the barrier increases to 3.1 kcal/mol. In **1,2,4,6-OTMTB**, the two-fold symmetry is restored by introducing the *t*-butyl groups at both sides, but the hydroxyl group is surrounded by a very crowded environment. Although the in-plane conformation is still the ground state, the perpendicular conformation is not the transition state but an intermediate state. The rotational transition state occurs at the same position, τ(hydroxyl) = 60°, as in **1,2,4-OTMB**. The rotational barrier decreases to only 0.5 kcal/mol. Presumably, the barrier is low because the ground state energy is elevated due to the presence of the two *t*-butyl groups. To further investigate the effect of the *t*-butyl groups to the hydroxyl rotation in **1,2,4,6-OTMTB**, the potential energy surfaces for the hydroxyl rotation were calculated at three fixed *t*-butyl orientations, both-out, both-in and in-out. The conformations are drawn in Scheme 4.3 and the PESs are shown in Figure 4.11 by the solid symbols. The orientation 'in' refers to the orientation of the *t*-butyl group with one of the Cm-Ct bonds coplanar with the benzene ring on the C1 side. The orientation 'out' refers to the coplanar Cm-Ct bond on the C5 or C3 side. See Scheme 4.3.

Due to the imbalance of the interactions from the two *t*-butyl groups in the in/out orientation, the hydroxyl group has the highest rotational barrier among the three orientations. For the both-in orientation, the perpendicular hydroxyl conformation is the transition state and the ground state is at τ(hydroxyl) = 30°. The barrier is about 0.4 kcal/mol.

It is tempting to divide the hydroxyl rotational barrier among the intramolecular interactions from the two *t*-butyl groups and the benzene ring. The interaction from the *t*-butyl groups is a repulsive steric effect; the interaction from the benzene ring is hyperconjugation. Are the interactions in **1,2,4,6-OTMTB** comparable to those in **1,2,4-OTMB**? The same question may be asked for 4-methylphenol (**1,4-OMB**) and phenol(**OB**) with regard to the hyperconjugation effect.

To answer the questions, we construct the PESs of **1,2,4,6-OTMTB** from the computed PESs in the isolated molecules of **1,2,4-OTMB** and **1,4-OMB** for the hydroxyl group rotation. We call them as 'synthetic' PESs. The following equation represents the construction scheme.

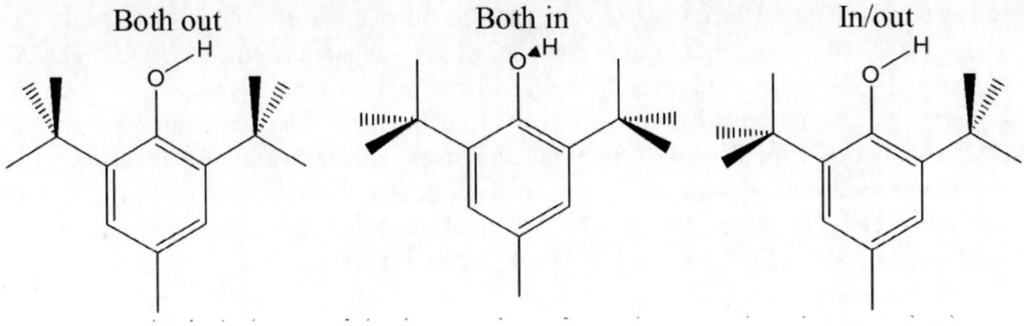

Scheme 4.3.

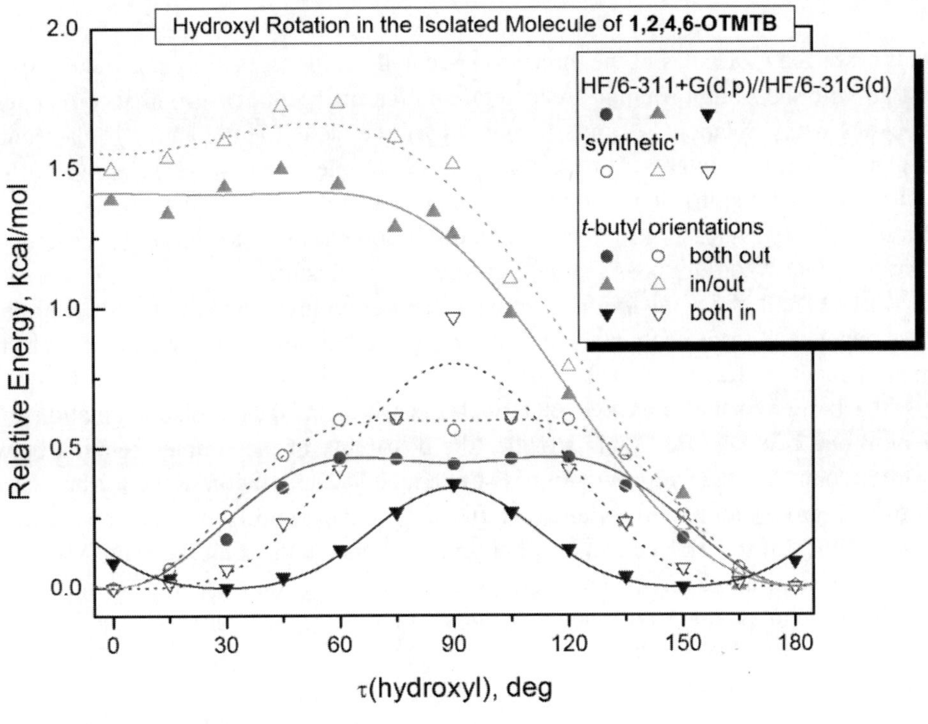

Figure 4.11. The rotational potential energy surfaces for the hydroxyl group in the isolated molecule of 2,6-di-*t*-butyl-4-methylphenol with the *t*-butyl groups fixed at three special orientations in Scheme 4.3. The solid symbols are the calculated results at the HF/6-311+G(d,p)//HF/6-31G(d) level. The open symbols are the 'synthetic' values from the potential energy surfaces of 2-*t*-butyl-4-methylphenol and 4-methylphenol. See the text for the construction scheme of the synthetic values.

E(**1,2,4,6-OTMTB**, τ_{OH}, $\tau_{2\text{-}t\text{-butyl}}$, $\tau_{6\text{-}t\text{-butyl}}$) = E(**1,2,4-OTMB**, τ_{OH}, $\tau_{2\text{-}t\text{-butyl}}$) + E(**1,2,4-OTMB**, τ_{OH}, $\tau_{6\text{-}t\text{-butyl}}$) – E(**1,4-OMB**, τ_{OH}).

In the above equation, E is the energy of the specified compound relative to its ground state. Since the hyperconjugation between the hydroxyl group and the benzene ring is counted twice in the first two terms on the right-hand side of the equation, the third term is required to subtract the effect. In Figure 4.11, the three 'synthetic' curves are shown by open symbols. The synthetic PESs reproduce well the shape of the computed PESs. The barriers found in this way are also close to the computed values for the orientations of both-out and in-out. The exception is the barrier in the both-in orientation, which is higher than the computed value by 0.6 kcal/mol. The success of this synthetic scheme shows that we may approximately decompose the intramolecular interactions involved in the hydroxyl rotation into several components between the hydroxyl group and fragments of the rest of the molecules. The interactions are nearly the same across the compounds. The out-lying barrier in the both-in orientation can be explained by that the structural relaxation is slightly different in the molecules of **1,2,4,6-OTMTB** and **1,2,4-OTMB** and other factors.

4.5.4. Rotational Coupling between the *t*-Butyl Group, the Methyl Groups and the Hydroxyl Group

In the isolated molecules of the three aromatic hydrocarbons, the in-plane methyl rotation is coupled with the *t*-butyl rotation. A 5° reorientation of the *t*-butyl group accompanies the out-of-plane methyl rotation. A similar magnitude of the methyl reorientation is also found to accompany the *t*-butyl rotation. In the phenols, the coupled rotation route between the in-plane methyl group and the *t*-butyl group becomes unfavorable due to the high barrier to the *t*-butyl rotation. However, the geared reorientation of some other rotor groups does happen in the rotation of one rotor group, as shown in the section of Results.

In **1,2,4-OTMB**, the paths for the rotation of the *t*-butyl group and the hydroxyl group are nearly independent from each other except the geared reorientation within 10°. But the situation changes in **1,2,4,6-OTMTB**. Because of the steric crowding from the two *t*-butyl groups, the barrier to the hydroxyl rotation decreases to 0.5 kcal/mol; the rotation of the hydroxyl group is almost free. This provides the possibility of the coupled rotation between the *t*-butyl groups and the hydroxyl group. For example, in the rotation of the 6-*t*-butyl group, the hydroxyl group adjusts its orientation by 45° in the optimized structures. The two-dimensional PES of the molecule shows that the rotational path of the 6-*t*-butyl group has a lower barrier than that of the 2-*t*-butyl group. There is not much difficulty for the 2-*t*-butyl group converting to the 6-*t*-butyl group through the rotation of the hydroxyl group.

4.6. CONCLUSION

The internal rotations in the isolated molecules of two phenol compounds have been studied. The potential energy surfaces for the internal rotation of the *t*-butyl groups, the methyl groups and the hydroxyl groups are computed at the HF/6-311 +G(d,p)//HF/6-31G(d) level. From the above results and discussion, the following conclusions can be drawn.

(1) The presence of the ortho hydroxyl group causes a strong hindrance to the *t*-butyl rotation. The barrier to the 2-*t*-butyl rotation is 3.58 kcal/mol in 2-*t*-butyl-4-methylphenol. The barriers to the 2-*t*-butyl rotation and to the 6-*t*-butyl rotation are

4.08 and 3.09 kcal/mol, respectively, in 2,6-di-*t*-butyl-4-methyphenol. The ground state orientation has one of the Cm-Ct bonds coplanar with the benzene ring on the farther side away from the hydroxyl group.

(2) The internal rotation of the methyl groups is affected slightly by the hydroxyl group: the barrier to the out-of-plane methyl rotation is smaller than in the aromatic hydrocarbons reported in Chapter 2; the barrier to the in-plane methyl rotation increases slightly. The barrier to the out-of-plane methyl rotation is 3.36 kcal/mol in 2-*t*-butyl-4-methylphenol, 3.06 kcal/mol for the 2-*t*-butyl group and 3.25 kcal/mol for the 6-*t*-butyl group in 2,6-di-*t*-butyl-4-methylphenol. The barrier to the in-plane methyl rotation is 7.34 kcal/mol in 2-*t*-butyl-4-methylphenol, 7.66 kcal/mol for the 2-*t*-butyl group and 7.69 kcal/mol for the 6-*t*-butyl group in 2,6-di-*t*-butyl-4-methylphenol. The ground state orientation for all the methyl groups is the staggered conformation, τ(methyl) = 60°.

(3) The barrier to the hydroxyl group rotation increases to 3.11 kcal/mol in 2-*t*-butyl-4-methylphenol, compared to 2.5 kcal/mol in phenol itself. The barrier is only 0.46 kcal/mol in 2,6-di-*t*-butyl-4-methylphenol.

(4) The internal rotations of the *t*-butyl groups and the methyl groups are largely independent. But the geared reorientation in the range from 5 to 10° of one group is necessary for the rotation of the other. The internal rotations of the *t*-butyl groups and the hydroxyl groups are largely independent in 2-*t*-butyl-4-methylphenol except the geared reorientation within 10°. But the coupling between the internal rotations of the *t*-butyl groups and the hydroxyl group occurs in 2,6-di-*t*-butyl-4-methylphenol due to the nearly free rotation of the hydroxyl group.

(5) The ortho hydroxyl group affects the *t*-butyl group rotation in two ways: hyperconjugation between the hydroxyl group and the benzene ring, and the steric repulsion between the *t*-butyl groups and the hydroxyl group. Both effects contribute to the three-fold component for the PES of the *t*-butyl rotation. The contribution from the hyperconjugation is estimated to be less than 0.5 kcal/mol, and the latter effect is a dominant factor for the internal rotation of the *t*-butyl groups.

4.7. REFERENCES

[1] Berg, U; Liljefors, T; Roussel, C; Sandstrom, J. *Acc. Chem. Res.*, 1985, 18, 80.

[2] Jackman, LM; Cotton, FA. *Dynamic NMR Spectroscopy*. Academic Press: New York, 1975.

[3] Oki, M. *Applications of Dynamic NMR Spectroscopy to Organic Chemistry*. VCH Publishers: Deerfield Beach, FL, 2005.

[4] Seeman, JI; Secor, HV; Breen, PJ; Grassian, VH; Bernstein, ER. A study of nonrigid aromatic molecules. Observation and spectroscopic analysis of the stable conformations of various alkylbenzenes by supersonic molecular jet laser spectroscopy. *J. Am. Chem. Soc.*, 1989, 111, 3140-3150.

[5] Beckmann, PA; Buser, CA; Gullifer, K; Mallory, FB; Mallory, CW; Rossi, GM; Rheingold, AL. Methyl and *t*-butyl group reorientation in planar aromatic solids: Low-

frequency nuclear magnetic resonance relaxometry and X-ray diffraction. *J. Chem. Phys.*, 2003, 118(24), 11129-11138.

[6] Beckmann, PA; Fusco, FA; O'Neill, A. E. Proton spin-lattice relaxation and intramolecular reorientation in solids. *J. Magn. Reson.*, 1984, 59, 63-70.

[7] Beckmann, PA; Paty, C; Allocco, E; Herd, M; Kuranz, C; Rheingold, AL. The relationship between crystal structure and methyl and *t*-butyl group dynamics in van der Waals organic solids. *J. Chem. Phys.*, 2004, 120(11), 5309-5314.

[8] Beckmann, PA; Ratcliffe, CI; Dunell, B. Proton spin-lattice relaxation and methyl group rotation. *J. Magn. Reson.*, 1978, 32, 391-402.

[9] Richardson, PR; Chapman, MA; Wilson, DC; Bates, SP; Jones, AC. The nature of conformational preference in a number of *p*-alkylphenols and *p*-alkyl benzenes. *Phys. Chem. Chem. Phys.*, 2002, 4, 4910-4915.

[10] Zierkiewicz, W; Michalska, D; Hobza, P. The barrier to internal rotation and electronic effects in *para*-halogenophenols: theoretical study. *Chem. Phys. Lett.*, 2004, 386, 95-100.

[11] Im, H.-S; Bernstein, ER; Secor, HV; I, SJ. Supersonic jet studies of benzyl alcohols: Minimum energy conformations and torsional motion. *J. Am. Chem. Soc.*, 1991, 113, 4422-4431.

[12] Parr, WJE; Schaefer, T. A proton magnetic resonance determination of the barrier to internal rotation in benzethiol in solution. *J. Magn. Reson.,* 1977, 25, 171-176.

[13] Pedersen, T; Larsen, NW; Nygaard, L. *J. Mol. Struct.*, 1969, 4, 59.

[14] Mathier, E; Welti, D; Bauder, A; Günthard, Hs. H. *J. Mol. Spectr.* 1971, 37, 63.

[15] Larsen, NW; Nicolaisen, FM. *J. Mol. Struct.*, 1974, 22, 29.

[16] Yamamoto, G; Oki, M. *Tetrahedron Lett.*, 1986, 49.

[17] Yamamoto, G; Oki, M. *Bull. Chem. Soc., Jpn.*, 1986, 59, 3597.

[18] Horsewill, AJ. Quantum tunneling aspects of methyl group rotation studied by NMR. *Prog. Nucl. Magn. Reson. Spectroscopy* ,1999, 35, 359-389.

[19] Spangler, LH. Structural information from methyl internal rotation spectroscopy. *Ann. Rev. Phys. Chem.*, 1997, 48, 481-510.

[20] Lu, K.-T; Richard, EC; Walker, RA; Weisshaar, JC. Methyl internal rotation in substituted toluenes: Effects of substituents, electronic excitation, and ionization. *NATO ASI Series, Series C: Mathematical and Physical Sciences*, 1999, 521, 307-334.

[21] Schiebel, P; Amoretti, G; Ferrero, C; Paci, B; Prager, M; Caciuffo, R. Coherent internal and centre of mass quantum rotation of methyl groups in the *p-tert*-butyl-calix[4]arene (2:1) *p*-xylene supramolecular complex. *J. Phys.: Condens. Matter*, 1998, 10, 2221-2231.

[22] Sonoda, Y; Iwata, S. Theoretical studies of the internal rotation of the methyl group in *o*-, *m*-, and *p*-fluorotoluenes and their cations. *Chem. Phys. Lett.*, 1995, 243(1/2), 176-182.

[23] Frisch, MJ; Trucks, GW; Schlegel, HB; Scuseria, GE; Robb, MA; Cheeseman, JR; Montgomery, Jr., JA; Vreven, T; Kudin, KN; Burant JC; Millam, JM; Iyengar, SS; Tomasi, J; Barone, V; Mennucci, B; Cossi, M; Scalmani, G; Rega, N; Petersson, GA; Nakatsuji, H; Hada, M; Ehara, M; Toyota, K; Fukuda, R; Hasegawa, J; Ishida, M; Nakajima, T; Honda, Y; Kitao, O; Nakai, H; Klene, M; Li, X; Knox, JE; Hratchian, HP; Cross, JB; Bakken, V; Adamo, C; Jaramillo, J; Gomperts, R; Stratmann, RE; Yazyev, O; Austin, AJ; Cammi, R; Pomelli, C; Ochterski, JW; Ayala, PY; Morokuma, K; Voth,

GA; Salvador, P; Dannenberg, JJ; Zakrzewski, VG; Dapprich, S; Daniels, AD; Strain, MC; Farkas, O; Malick, DK; Rabuck, AD; Raghavachari, K; Foresman, JB.; Ortiz, JV; Cui, Q; Baboul, AG; Clifford, S; Cioslowski, J; Stefanov, BB; Liu, G; Liashenko, A; Piskorz, P; Komaromi, I; Martin, RL; Fox, DJ; Keith, T; Al-Laham, MA; Peng, CY; Nanayakkara, A; Challacombe, M; Gill, P. MW; Johnson, B; Chen, W; Wong, MW; Gonzalez, C; Pople, JA. *Gaussian 03, Revision B.05*, Gaussian, Inc., Pittsburgh PA, 2003.

[24] Glendening, ED; Reed, AE; Carpenter, JE; Weinhold F. *NBO Version 3.1*.

[25] The X-ray crystal structures are available free of charge from the Cambridge Crystallographic Data Centre. See http://www.ccdc.cam.ac.uk/ for details.

[26] Bolte, M; Amon, M. *Private Communication*, 2000.

[27] Hoffmann, R; Pople, JA; Radom, L; Schleyer, PvR; Hehre, WJ; Salem, L. Strong conformational consequences of hyperconjugation. *J. Am. Chem. Soc.*, 1972, 94(17), 6221-6222.

[28] Pophristic, V; Goodman, L. Hyperconjugation not steric repulsion leads to the staggered structure of ethane. *Nature*, 2001, 411, 565-568.

[29] Sadlej-Sosnowska, N. Energy Barriers to internal rotation: Hyperconjugation and electrostatic description. *J. Phys. Chem. Am*, 2003, 107, 8671-8676.

[30] Reed, AE; Weinhold, F. *J. Chem. Phys.*, 1983, 78, 4066.

[31] Reed, AE; Weinhold, F. *J. Chem. Phys.*, 1985, 83, 1736.

[32] Reed, AE; Weinstock, RB; Weinhold, F. *J. Chem. Phys.*, 1985, 83, 735.

[33] Reed, AE; Curtiss, LA; Weinhold, F. *Chem. Rev.*, 1988, 88, 899.

[34] Weinhold, F; Carpenter, JE. *Plenum* 1988, 227.

[35] Carpenter, JE; Weinhold, F. *J. Mol. Struct.* (*Theochem*), 1988, 169, 41.

[36] Carpenter, JE. *PhD thesis*, University of Wisconsin, Madison, WI, 1987.

[37] Foster, JP; Weinhold, F. *J. Am. Chem. Soc.*, 1980, 102, 7211.

Chapter 5

INTERNAL ROTATIONS OF THE PHENOLS IN THE CRYSTALLINE SOLID STATE

5.1. ABSTRACT

Internal rotations of 2-*t*-butyl-4-methylphenol (**1,2,4-OTMB**) and 2,6-di-*t*-butyl-4-methylphenol (**1,2,4,6-OTMTB**) in their crystalline solids are studied using the cluster model. The potential energy surfaces of the rigid rotation and the partially relaxed rotation of the *t*-butyl groups, the constituent methyl groups, and/or the hydroxyl groups were calculated in the clusters at the HF/3-21G level and the B3LYP/6-31G(d) level. The barrier to the partially relaxed rotation is 12.9 kcal/mol for the 2-*t*-butyl group at the crystallographic site 1 of **1,2,4-OTMB**, 12.5 kcal/mol for the 2-*t*-butyl group at the site 2 of **1,2,4-OTMB**, 11.6 kcal/mol for the 2-*t*-butyl group at the unique site of **1,2,4,6-OTMTB**, and 9.49 kcal/mol for the 6-*t*-butyl group of **1,2,4,6-OTMTB**, computed at the B3LYP/6-31G(d) level. The barrier to the rigid rotation of the in-plane or nearly in-plane methyl groups is about 7~8 kcal/mol, and the barrier to the rigid rotation of the out-of-plane methyl groups is approximately 4 kcal/mol. The barrier to the rigid rotation of the hydroxyl groups is 11.5 kcal/mol in the clusters of **1,2,4-OTMB**, and the barrier is 6.34 kcal/mol in the clusters of **1,2,4,6-OTMTB**. The computation method is also capable of decomposing the barriers into the contributions from neighboring molecules and from the intramolecular interactions. The results provide helpful and complementary information in interpreting the proton spin-lattice relaxation rate data measured in low frequency solid state nuclear magnetic resonance relaxometry.

5.2. INTRODUCTION

In this chapter, the results of the computational study on the internal rotations of the two phenol compounds, 2-*t*-butyl-4-methylphenol (**1,2,4-OTMB**) and 2,6-di-*t*-butyl-4-methylphenol (**1,2,4,6-OTMTB**), in the crystalline solid state are presented. The computational techniques that have been established in Chapter 3, including the cluster model based on the X-ray crystal structure and the potential energy surfaces (PESs) of the rigid rotation and the partially relaxed rotation, were adopted for the study.

Solid state proton nuclear magnetic resonance (NMR) relaxometry has been used to study the internal rotation dynamics of a series of phenols at various Larmor frequencies from 8.5 to

59 MHz [1-5]. By measuring the temperature dependence of the proton spin-lattice relaxation rates ($1/T_1$) in polycrystalline samples, one is able to extract information on the internal rotations of *t*-butyl groups and methyl groups, which modulate the spin-dipolar interactions among the protons. The proton relaxation rate curves of **1,2,4,6-OTMTB** below the temperature of 80 K are dominated by the motions of the 4-methyl group. The barrier to the internal rotation of this group is lower than 1 kcal/mol, and the motion is not the subject of this study. Instead, we focus on the internal rotations of the *t*-butyl groups, their constituent methyl groups and the hydroxyl groups in the two phenols. The information of their motions is embedded in the relaxation data at high temperatures. The proton relaxation rate curves of both compounds show two well separated maxima at high temperatures which suggest a minimum of two distinct motions. However, a detailed analysis is complicated by the presence of multiple rotational degrees of freedom. In the crystal of **1,2,4,6-OTMTB**, the 2- and 6-*t*-butyl groups have different environments which may lead to significantly different barriers to the internal rotation of the *t*-butyl groups and their constituent methyl groups. Although there is only one *t*-butyl group in a molecule of **1,2,4-OTMB**, there are two inequivalent molecular sites in its crystal, and the *t*-butyl groups at the two sites are in different neighborhoods. In both compounds, we also need to consider how the hydroxyl groups at the ortho position relative to the *t*-butyl groups affect the motions of the *t*-butyl groups and the constituent methyl groups. Because of the complication, there are several dynamical models that can be fitted to the relaxation rate curves. Computational studies may provide a solution for this situation and give a satisfactory interpretation of the experimental results. Furthermore, we wish to learn how intramolecular and intermolecular interactions contribute to the barriers to the internal rotations in similar molecular systems by investigating the two compounds.

5.3. Methods

The cluster models were built in the algorithm described in Chapter 3 from the X-ray crystal structure of the compounds. More details can be found in the Results section.

Potential energy surfaces of the clusters were calculated at two theoretical levels, HF/3-21G and B3LYP/6-31G(d), using the *Gaussian03* suite [6] without correcting the basis-set superposition error as discussed in Chapter 3. Other details can also be found in the Methods section in Chapter 3.

Table 5.1. Crystal unit cell parameters of the two compounds, measured in the X-ray diffraction method

Parameter	2-*t*-butyl-4-methylphenol	2,6-di-*t*-butyl-4-methylphenol
code[a]	BECK04 [c][8-9]	MBPHOL02 [11]
space group	$P2_1/c$	$P2_12_12_1$
a, Å	6.2618(4)	8.7562(8)
b, Å	16.6852(10)	10.2435(9)

Parameter	2-*t*-butyl-4-methylphenol	2,6-di-*t*-butyl-4-methylphenol
c, Å	18.9674(12)	15.435(2)
α,°	90	90
β,°	94.1670(10)	90
γ,°	90	90
V, Å^3	1976.46	1384.43
Z	8	4
T, K[b]	173(2)	173(2)

[a] Entry code in the Cambridge Crystal Database.
[b] Measurement temperature.
[c] Private code.

5.4. RESULTS

5.4.1. Cluster Models

Crystal structures of **1,2,4-OTMB** and **1,2,4,6-OTMTB** were determined previously by X-ray crystallography and deposited in the Cambridge Crystal Database (CCDC) [7]. They were used to build the cluster models simulating *t*-butyl group rotation environment in the crystalline solids. The crystal of **1,2,4-OTMB** (private code: BECK04 [8-9]) belongs to the monoclinic space group of $P2_1/c$. Two polymorphs of **1,2,4,6-OTMTB** have been found (CCDC refcode: MBPHOL01 [10], MBPHOL02 [11] and MBPHOL10 [12]). The polymorph of MBPHOL01 belongs to the monoclinic space group of *C*2/*c* with eight molecules in its unit cell. The other polymorph of MBPHOL02 and MBPHOL10, which is studied here, belongs to the orthorhombic space group of $P2_12_12_1$. The more recently determined structure of MBPHOL02 was used to build the cluster. The unit cell parameters of BECK04 and MBPHOL02 are listed in Table 5.1. Both structures were determined at 173 K.

There are two unique molecules of **1,2,4-OTMB** in the crystal of BECK04. At one molecular site, the *t*-butyl group has its thermal equilibrium orientation at τ(*t*-butyl) = -9.2°, and at the other site, τ(*t*-butyl) = 1.7°. Here and after, we will use the same definition of the internal rotation coordinates as is defined for the isolated molecules in the previous chapter. The two molecular sites will be referred as site 1 and site 2, respectively. Two distinct clusters were built to mimic the crystal environment for the *t*-butyl group at the two sites. The cluster of the site 1 includes 13 molecules of **1,2,4-OTMB**, of which the 12 neighboring molecules have at least one atom falling into an "interaction sphere" centered at the Ct carbon atom of the central *t*-butyl group with a radius of 7.1 Å. The cluster of the site 2 includes 14 molecules, and the threshold radius of the interaction sphere is 6.8 Å.

There is only one unique molecule of **1,2,4,6-OTMTB** in the crystal of MBPHOL02. However, there are two *t*-butyl groups on a molecule of **1,2,4,6-OTMTB**. Not only are the two *t*-butyl groups not identical in terms of their intramolecular environments, but also they have quite different intermolecular neighborhoods. They will be arbitrarily distinguished as 2-*t*-butyl and 6-*t*-butyl groups hereafter. The 2-*t*-butyl group refers to the one with τ(*t*-butyl) = 1.0°, and the 6-*t*-butyl group has τ(*t*-butyl) = -5.3°. Although with one big cluster it would be possible to accommodate the necessary neighboring molecules of both *t*-butyl groups, we built two separate smaller clusters for them to avoid the impractically expensive computational cost for a cluster that is too large. The critical radius was chosen at 6.5 Å for the 2-*t*-butyl group cluster and at 7.3 Å for the 6-*t*-butyl group cluster. There are 13 molecules in each of these clusters.

Perspective structures of the clusters are drawn in Figure 5.1. For the convenience of description, the molecules and/or the *t*-butyl groups are labeled in letters and/or numbers. The central *t*-butyl group is the one on the molecule 'a' in the cluster of **A** and on the molecule 'b' in the cluster of **B**. In the clusters of **C** and **D**, the central *t*-butyl group is the one labeled by '1' and '2', respectively.

A. 1,2,4-OTMB (site 1)

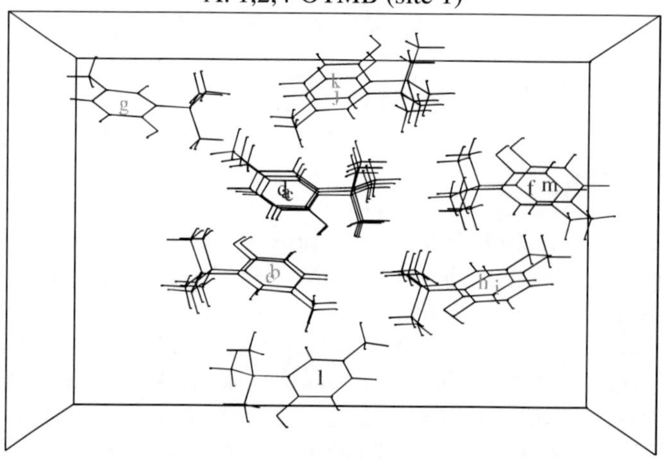

B. 1,2,4-OTMB (site 2)

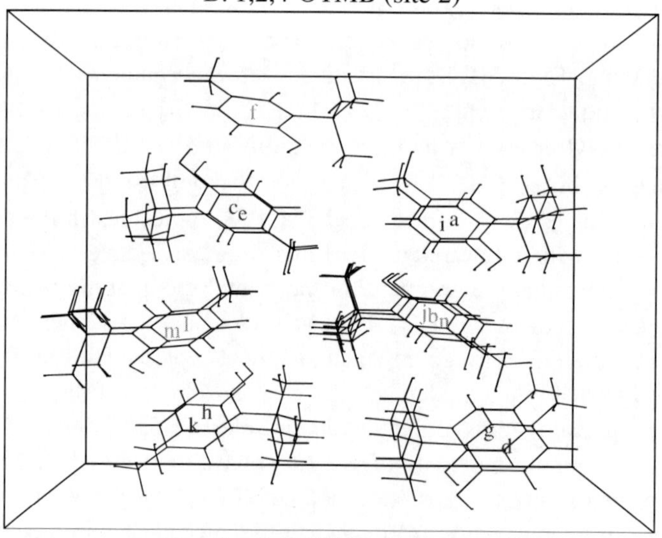

C. 1,2,4,6-OTMTB (2-*t*-butyl)

Figure 5.1. (Continued).

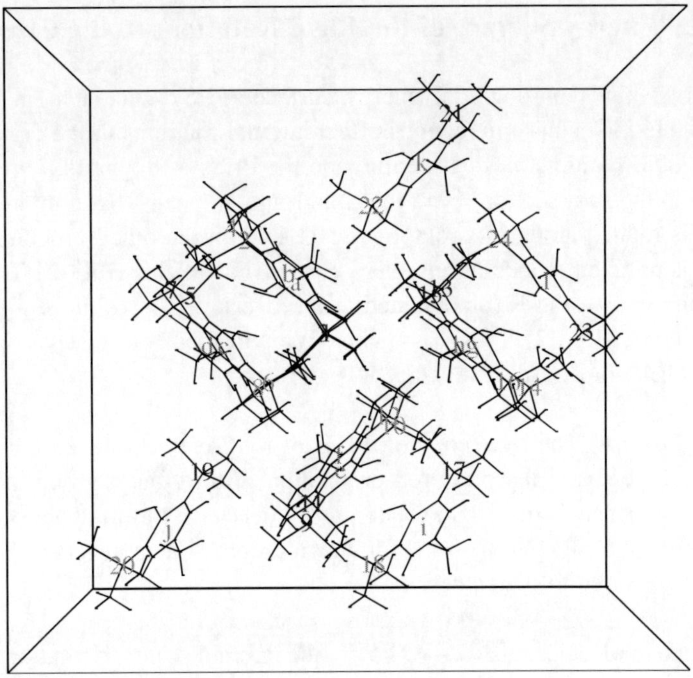

D. 1,2,4,6-OTMTB (6-*t*-butyl)

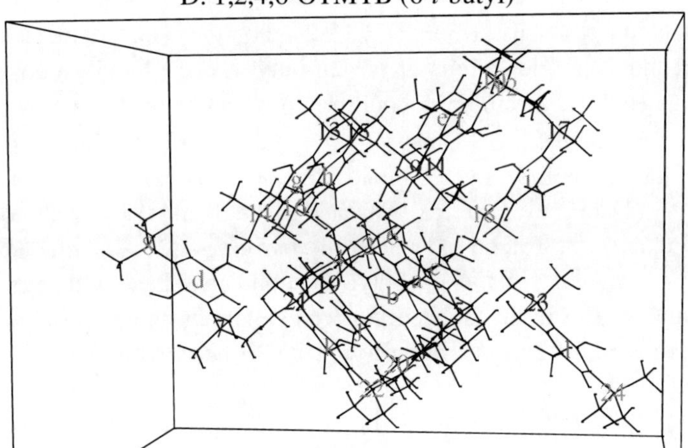

Figure 5.1. Structure of the cluster models of 2-*t*-butyl-4-methylphenol (A and B) and 2,6-di-*t*-butyl-4-methylphenol (**C** and **D**). The molecules and/or *t*-butyl groups are labeled in an arbitrary order. Blue and green colors stand for site 1 and site 2, respectively, in A and B, and for 2-*t*-butyl and 6-*t*-butyl groups, respectively, in **C** and **D**. The central *t*-butyl group is the one labeled by 'a' in **A**, 'b' in **B**, '1' in **C**, and '2' in **D**. The frames serve as a guide to the eyes. The outer boxes are closer to the reader.

5.4.2. Potential Energy Surface of the Rigid Rotations in the Clusters

There are two kinds of potential energy surface (PES) calculated in the clusters to investigate how intermolecular interactions affect internal rotations in the crystals. We report the PESs of the rigid rotations in this section, and the PESs of the partially relaxed rotations will be reported in the next section. Here the rigid rotation means that only the orientation of the rotor group is rotated around its rotational axis, and all the other structure parameters of the cluster are kept fixed. Calculations were carried out at the HF/3-21G and B3LYP/6-31G(d) levels. In general, the PESs obtained at the two levels have the same trend, but the barrier obtained at the HF/3-21G level is higher. We will focus our discussion on the results obtained at the B3LYP/6-31G(d) level since it is more accurate.

(a) ***t*-Butyl Groups**. The rigid rotation PESs of the *t*-butyl groups are shown in Figure 5.2. In all the cases, the preferred orientation of the *t*-butyl group in the crystalline solid is confirmed as the ground state in the clusters. By turning about 60° around the Ca-Ct bond, the transition state for the rotation is then reached. The PESs have approximate three-fold symmetry.

In the isolated molecules of **1,2,4-OTMB**, the ground state orientation of the *t*-butyl group is τ(*t*-butyl) = 0°. This conformation is very similar to the thermal equilibrium orientation found at the site 2 in the crystal, τ(*t*-butyl) = 1.7°. But at the site 1, the thermal equilibrium orientation shifts to τ(*t*-butyl) = -9.2°. For the isolated molecules of **1,2,4,6-OTMTB**, the ground state orientations of the 2-*t*-butyl and 6-*t*-butyl groups are both 0°. In the crystal of MBPHOL02, the thermal equilibrium orientation of the two groups is τ(2-*t*-butyl) = 1.0° and τ(6-*t*-butyl) = -5.3°, respectively. In contrast to the barrier of 3.58 kcal/mol for the *t*-butyl rotation in the isolated molecules of **1,2,4-OTMB**, the barrier increases dramatically in the clusters to 21.0 kcal/mol at the site 1 and 19.4 kcal/mol at the site 2. Similarly, in the clusters of **1,2,4,6-OTMTB**, the barrier is 22.3 kcal/mol for the 2-*t*-butyl rotation, and 17.7 kcal/mol for the 6-*t*-butyl rotation, compared with the 4.08 and 3.00 kcal/mol for the two groups in the isolated molecules. Furthermore, not only does the barrier increases significantly in the clusters, but also for the 2-*t*-butyl group and the 6-*t*-butyl group are different in the clusters.

The PESs of the *t*-butyl rotation in the isolated molecules can be fitted to $V(\tau) = \frac{V_3}{2}(1-\cos 3\tau) + \frac{V_6}{2}(1-\cos 6\tau)$. A similar model can also be used to fit the PESs obtained in the clusters. But a two-fold cosine term is needed to account the energy difference between the conformation at τ(*t*-butyl) = 120° and the ground state.(This is an artifact effect for the rigid rotation PES and the difference should disappear if the *t*-butyl groups are allowed to relax.) In addition, the phase of the 3-fold cosine term is also shifted according to the ground state orientation. The least-square best fit curves are plotted in the dotted and solid lines in Figure 5.2, and the best fit coefficients of the model, $V(\tau) = \frac{V_2}{2}(1-\cos 2\tau) + \frac{V_3}{2}(1-\cos 3(\tau-\tau_0)) + \frac{V_6}{2}(1-\cos 6\tau)$, are collected in Table 5.2. (Due to the limited number of data points, there are large errors in some fitting results.)

A. 1,2,4-OTMB

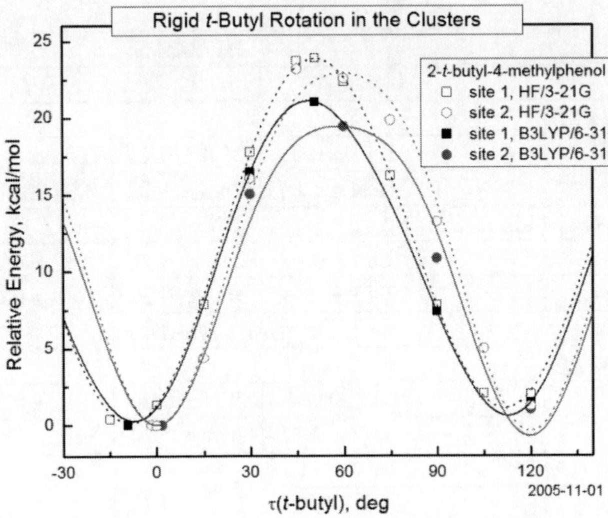

B. 1,2,4,6-OTMTB

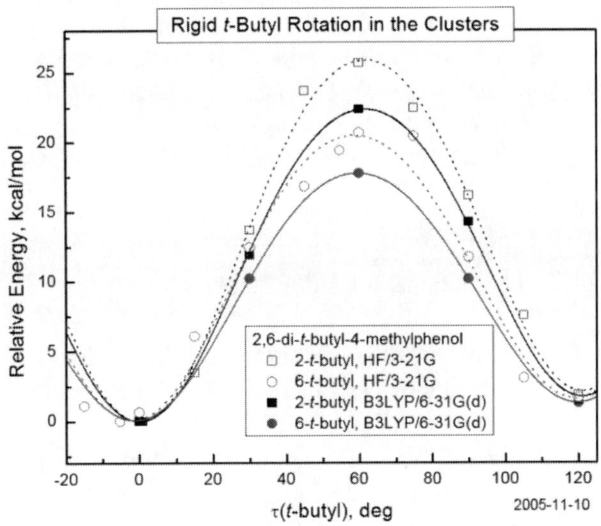

Figure 5.2. The potential energy surfaces of the rigid rotation of the *t*-butyl groups in the clusters computed at the HF/3-21G and B3LYP/6-31G(d) levels. The solid and dotted curves are the best-fit results with a sum of the two-fold, three-fold and six-fold cosine potentials. The best-fit coefficients are listed in Table 5.2.

Table 5.2. The least-square best fit results for the rigid rotation potential energy surfaces of the *t*-butyl groups in the clusters

Components, kcal/mol	HF/3-21G		
	V_2	V_3	V_6
1,2,4-OTMB			
site 1	1.17 ± 0.22	23.0 ± 0.21	0.42 ± 0.23
site 2	-0.54 ± 1.07	23.3 ± 1.22	3.37 ± 1.07
1,2,4,6-OTMTB			
2-*t*-butyl	2.85 ± 0.67	23.7 ± 0.76	1.32 ± 0.67
6-*t*-butyl	2.73 ± 0.69	16.5 ± 0.75	1.89 ± 0.68
	B3LYP/6-31G(d)		
1,2,4-OTMB			
site 1	0.43 ± 0.24	20.4 ± 0.26	1.40 ± 0.25
site 2	-0.97 ± 2.31	20.1 ± 2.74	3.48 ± 1.93
1,2,4,6-OTMTB			
2-*t*-butyl	2.37 ± 0.32	20.5 ± 0.38	1.33 ± 0.27
6-*t*-butyl	3.38 ± 1.48	15.0 ± 1.68	0.47 ± 1.39

(b) **Constituent Methyl Groups of the *t*-Butyl Groups**. The PESs for the rigid rotation of the in-plane or nearly in-plane methyl groups and the out-of-plane methyl groups on the central *t*-butyl groups are shown in Figures 5.3 and 5.4, respectively. Similar to the methyl group rotations in the isolated molecules, the PESs can be fitted into the equation, $V(\tau)=\frac{V_3}{2}(1-\cos 3(\tau-\frac{\pi}{3}))+\frac{V_6}{2}(1-\cos 6\tau)$. The least-square best fit coefficients are collected in Tables 5.3 and 5.4. Generally, the staggered conformation, τ(methyl) = 60°, is the ground state, and the eclipsed conformation, τ(methyl) = 0°, is the rotational transition state.

The in-plane methyl groups at the two sites of **1,2,4-OTMB** have nearly identical rotational barriers about 8.8 kcal/mol, and there is slight difference in their shapes. The difference is embodied in the six-fold components. The PESs of the two out-of-plane methyl groups at the site 1 have a larger difference. The rotational barrier for the out-of-plane methyl group labeled by τ(*t*-butyl) = 110.0° is 1.74 kcal/mol higher than the rotational barrier for the other out-of-plane methyl group labeled by τ(*t*-butyl) = -128.4°. This result is contrary to the prediction from the dependence of the methyl rotation barrier on the orientation of the *t*-butyl group, as studied in the previous chapter for the isolated molecules. (Because of the stronger hindrance for the latter methyl group caused by the interaction between the methyl group and the benzene ring, the latter methyl group should have a higher barrier. See the schematic drawing below, where the benzene ring is represented by a bold horizontal line.) Presumably, the higher barrier for the former methyl group is a result of stronger intermolecular interactions.

A. 1,2,4-OTMB

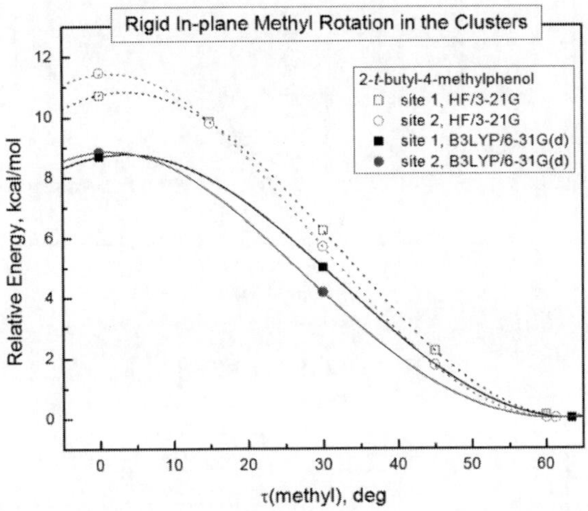

B. 1,2,4,6-OTMTB

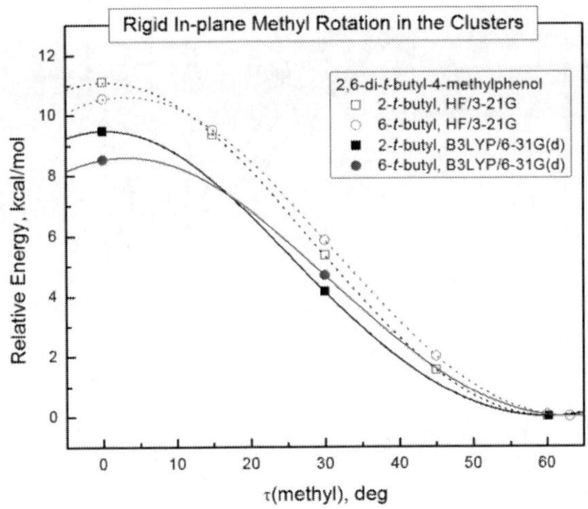

Figure 5.3. The potential energy surfaces of the rigid rotation of the in-plane or nearly in-plane methyl groups in the clusters computed at the HF/3-21G and B3LYP/6-31G(d) levels. The solid and dotted curves are the best-fit results with a sum of the three-fold and six-fold cosine potentials. The best-fit coefficients are listed in Table 5.3.

A. 1,2,4-OTMB

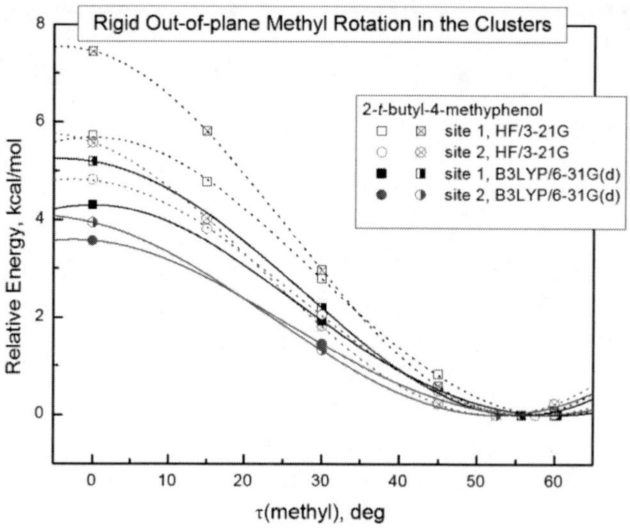

B. 1,2,4,6-OTMTB

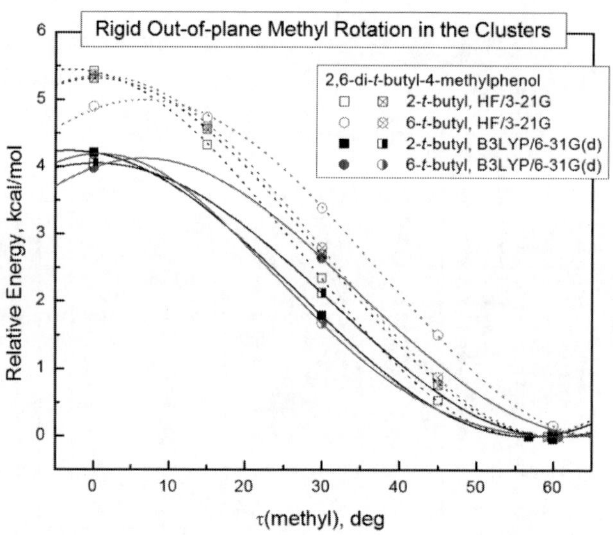

Figure 5.4. The potential energy surfaces of the rigid rotation of the out-of-plane methyl groups in the clusters computed at the HF/3-21G and B3LYP/6-31G(d) levels. The solid and dotted curves are the best-fit results with the three-fold and six-fold cosine potentials. The best-fit coefficients are listed in Table 5.4.

At the site 2, the *t*-butyl group gets closer to being coplanar to the ring. So the two out-of-plane methyls have nearly the same distance from the ring. The difference between the barriers to the rotation of the two out-of-plane methyl groups is smaller. One is 3.57 kcal/mol, and the other is 3.94 kcal/mol.

Table 5.3. The least-square best fit results for the rigid rotation potential energy surfaces of the in-plane or nearly in-plane methyl groups in the clusters

Components, kcal/mol	HF/3-21G		B3LYP/6-31G(d)	
	V_3	V_6	V_3	V_6
1,2,4-OTMB				
site 1	10.8 ± 0.01	-0.16 ± 0.01	8.76 ± 0.00	-0.16 ± 0.00
site 2	11.4 ± 0.04	-0.41 ± 0.05	8.84 ± 0.01	-0.56 ± 0.01
1,2,4,6-OTMTB				
2-*t*-butyl	11.1 ± 0.02	-0.20 ± 0.02	9.48 ± 0.00	-0.67 ± 0.00
6-*t*-butyl	10.6 ± 0.01	-0.34 ± 0.01	8.59 ± 0.00	-0.32 ± 0.00

Table 5.4. The least-square best fit results for the rigid rotation potential energy surfaces of the out-of-plane methyl groups in the clusters

Components, kcal/mol	HF/3-21G		B3LYP/6-31G(d)	
	V_3	V_6	V_3	V_6
1,2,4-OTMB				
site 1	5.69 ± 0.03	-0.06 ± 0.03	4.30 ± 0.00	-0.22 ± 0.00
site 1	7.55 ± 0.00	0.04 ± 0.00	5.25 ± 0.00	0.16 ± 0.00
site 2	4.83 ± 0.02	-0.06 ± 0.02	3.58 ± 0.00	-0.10 ± 0.00
site 2	5.45 ± 0.16	-1.21 ± 0.19	3.95 ± 0.05	-0.80 ± 0.05
1,2,4,6-OTMTB				
2-*t*-butyl	5.44 ± 0.01	0.07 ± 0.02	4.24 ± 0.05	0.03 ± 0.06
2-*t*-butyl	5.31 ± 0.02	0.03 ± 0.02	4.04 ± 0.00	0.10 ± 0.00
6-*t*-butyl	4.99 ± 0.02	0.06 ± 0.03	4.12 ± 0.01	-0.12 ± 0.02
6-*t*-butyl	5.38 ± 0.01	0.12 ± 0.02	4.18 ± 0.00	-0.42 ± 0.00

The four out-of-plane methyl groups on the 2- and 6-*t*-butyl groups in the clusters of **1,2,4,6-OTMTB** have a similar barrier of approximately 4.0 kcal/mol. The shapes of the PESs differ slightly, which is also embodied in the six-fold component coefficients. However, the barrier to the rotation of the in-plane methyl group on the 2-*t*-butyl group is 9.48 kcal/mol, 0.96 kcal/mol higher than the barrier to the rotation of the in-plane methyl group on the 6-*t*-butyl group. Furthermore, the order of the barriers is the same as the order of the *t*-butyl rotation barriers.

Compared with the barriers obtained in the isolated molecules, we find that the main origin of the barriers to the rotation of the methyl groups is from intramolecular interactions and intermolecular interactions only have a minor effect. The similar conclusion has been drawn for the rotation of the constituent methyl groups of the *t*-butyl groups on the aromatic hydrocarbons in the crystalline solid state (See Chapter 3). The out-of-plane methyl groups have a barrier near 4 kcal/mol and the in-plane methyl groups have a barrier about 7~8 kcal/mol.

(c) **Hydroxyl Groups**. The rigid rotation PESs of the hydroxyl groups on the central molecules in the clusters are also calculated at the HF/3-21G and B3LYP/6-31G(d) levels. They are shown in Figure 5.5.

In the isolated molecules of **1,2,4-OTMB**, the ground state orientation of the hydroxyl group is τ(OH) = 180°, in which the OH bond of the hydroxyl group is coplanar with the benzene ring and syn to the C1-C6 bond. The conformation at τ(OH) = 0° is a rotational intermediate state. This character is nearly preserved in the clusters. The PESs at the two molecular sites have a similar shape. In the clusters, the optimized orientation of the hydroxyl group is τ(OH) = 163° at the site 1 cluster and 167° at the site 2. The conformation at τ(OH) = 0° is the rotational intermediate. However, the barrier increases to 11.5 kcal/mol in the clusters, compared with 3.1 kcal/mol in the isolated molecules.

In the isolated molecules of **1,2,4,6-OTMTB**, the ground state orientation of the hydroxyl group is τ(OH) = 0° and the PES has a two-fold symmetry. The conformation at τ(OH) = 90° is the rotational intermediate. The rotational barrier is only about 0.5 kcal/mol. The PESs calculated in the two clusters are almost identical. This identity leads to the conclusion that both the clusters include enough neighboring molecules in the crystals which can possibly affect the rotation of the hydroxyl group. There is no apparent symmetry in the PESs. The optimized ground state orientation of the hydroxyl group is τ(OH) = 80° in the 2-*t*-butyl cluster and 82° in the 6-*t*-butyl cluster. Therefore, the preferred orientation of the hydroxyl group is approximately perpendicular (τ(OH) = 90°) to the benzene ring. In addition, the hydroxyl group points in a unique direction. (That is, the two faces of the benzene ring have different packing interactions in the clusters.) If the hydroxyl group is perpendicular to the ring on the other side, *i.e.* τ(OH) = -90° (= 270°), the conformation is nearly an intermediate state, which is close to the transition state at τ(OH) = 225° in energy. The barrier to the rotation of the hydroxyl group is about 6.34 kcal/mol in the clusters.

In the isolated molecules of **1,2,4-OTMB**, the PES of the hydroxyl group can be fitted to the equation, $V(\tau)=\frac{V_1}{2}(1-\cos(\tau-\pi))+\frac{V_2}{2}(1-\cos 2\tau)+\frac{V_4}{2}(1-\cos 4\tau)$. Due to the structural symmetry, the one-fold term is absent in **1,2,4,6-OTMTB**. Apparently, this model will not fit well with the PESs obtained in the clusters because there lack the symmetry properties in the PESs. However, if we shift the phase of the cosine terms to account the different ground state orientation in the clusters, the model can fit with the PESs in the clusters of both compounds as shown in Figure 5.5. The least-square best fit coefficients are reported in Table 5.5. The V_1 term, which is the largest component in all four cases, being affected by intermolecular interactions, increases dramatically in the clusters as compared in the isolated molecules. It determines the equilibrium orientation of the hydroxyl group in the crystals and dominates the barrier to its rotation.

Table 5.5. The least-square best fit result of the barrier components for the rigid rotation potential of the hydroxyl groups in the clusters

Components, kcal/mol	**HF/3-21G**		
	V_1	V_2	V_4
1,2,4-OTMB			
site 1	12.8 ± 0.27	2.13 ± 0.30	0.77 ± 0.30
site 2	14.90 ± 0.35	2.38 ± 0.38	0.56 ± 0.38
1,2,4,6-OTMTB			
2-*t*-butyl	9.61 ± 0.36	2.56 ± 0.34	0.71 ± 0.35

Components, kcal/mol	HF/3-21G		
	V_1	V_2	V_4
6-*t*-butyl	9.60 ± 0.48	2.37 ± 0.44	0.61 ± 0.44
	B3LYP/6-31G(d)		
1,2,4-OTMB			
site 1	9.80 ± 0.23	4.90 ± 0.33	0.16 ± 0.37
site 2	10.6 ± 0.50	4.70 ± 0.71	0.39 ± 0.81
1,2,4,6-OTMTB			
2-*t*-butyl	6.19 ± 0.81	-0.60 ± 0.51	0.46 ± 0.65
6-*t*-butyl	5.90 ± 0.77	-0.79 ± 0.66	0.86 ± 0.70

A. 1,2,4-OTMB

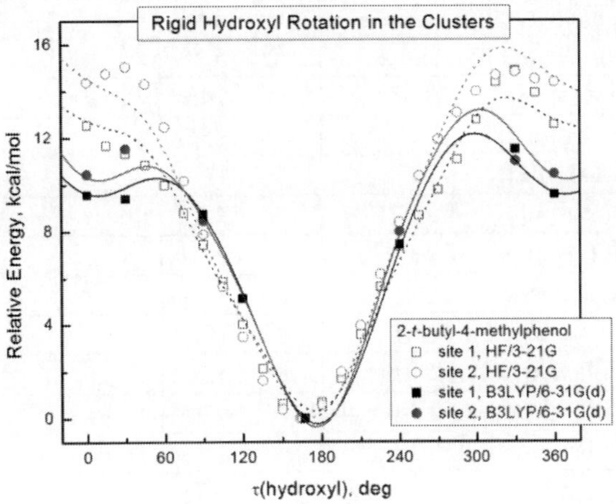

B. 1,2,4,6-OTMTB

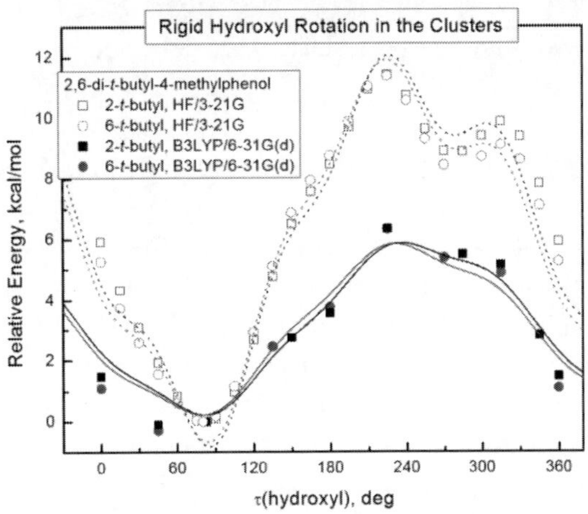

Figure 5.5. The potential energy surfaces of the rigid rotation of the hydroxyl groups in the clusters computed at the HF/3-21G and B3LYP/6-31G(d) levels. The solid and dotted curves are the best-fit results with a sum of the one-fold, two-fold and four-fold cosine potentials. The best-fit coefficients are listed in Table 5.5.

Table 5.6. The barriers to the rigid rotation of the *t*-butyl groups, the constituent methyl groups, and the hydroxyl groups in the clusters, calculated at the HF/3-21G level and the B3LYP/6-31G(d) level

	Barrier, kcal/mol	
	HF/3-21G	B3LYP/6-31G(d)
1,2,4-OTMB		
t-Butyl, site 1	23.8	21.0
t-Butyl, site 2	23.2	19.4
In-plane methyl, site 1	10.7	8.68
In-plane methyl, site 2	11.5	8.84
Out-of-plane methyl, site 1	5.72	4.30
Out-of-plane methyl, site 1	7.46	5.20
Out-of-plane methyl, site 2	4.81	3.57
Out-of-plane methyl, site 2	5.58	3.94
Hydroxyl, site 1	14.9	11.5
Hydroxyl, site 2	15.0	11.5
1,2,4,6-OTMTB		
t-Butyl, 2-*t*-butyl	25.7	22.3
t-Butyl, 6-*t*-butyl	20.7	17.7
In-plane methyl, 2-*t*-butyl	11.1	9.48
In-plane methyl, 6-*t*-butyl	10.5	8.52
Out-of-plane methyl, 2-*t*-butyl	5.41	4.21
Out-of-plane methyl, 2-*t*-butyl	5.35	4.05
Out-of-plane methyl, 6-*t*-butyl	4.86	3.98
Out-of-plane methyl, 6-*t*-butyl	5.35	4.19
Hydroxyl, 2-*t*-butyl cluster	11.4	6.34
Hydroxyl, 6-*t*-butyl cluster	11.4	6.34

The barriers to the rigid rotation in the clusters are collected in Table 6.

5.4.3. Potential Energy Surfaces of the Partially Relaxed Rotations

Calculations of the rigid rotations give an upper bound of the barrier to the internal rotations in the crystalline solids. However, it is rare that an internal rotation can be completely isolated from other modes. Couplings with other structural changes always accompany internal rotation, which may lower the barrier for large or small in both isolated molecules and clusters. We carried out calculations on the PES of the partially relaxed rotation in the clusters in order to investigate the effect of structural relaxation to internal rotation in the crystal solids. Particularly, we are interested in the possibility of geared rotations of *t*-butyl groups across neighboring molecules and intramolecular coupling between the rotation of the overall *t*-butyl group and the constituent methyl groups. For this purpose, two steps of partial optimization were done with the clusters at the rotational ground and transition states of the central *t*-butyl group. In the first step, all of the dihedrals of *t*-butyl groups (three of each) and hydroxyl groups in the clusters were optimized except the one labeling the orientation of the central *t*-butyl group. This optimization implies the freedom of the orientation of the *t*-butyl groups and the hydroxyl groups and the bond angles between the Ct-Cm bonds of the *t*-butyl groups. In the second step, structural parameters of the constituent methyl groups (bond lengths, bond angles and dihedrals) and the Ct atom of the central *t*-butyl group were optimized, besides the dihedrals of the *t*-butyl groups and the hydroxyl groups freed in the first step. This step includes the intramolecular relaxation effect along with the intermolecular relaxation from the last step. Due to the computational cost issues, the optimization was done at the HF/3-21G level and followed by a single point energy calculation at the B3LYP/6-31G(d) level. The use of the HF/3-21G level is justified by the fact that the model does predict the right shape of the PESs and thus the correct optimized structure. But the B3LYP/6-31G(d) level gives a much better estimation on the value of the barriers.

In Table 5.7, we list the barriers to the partially relaxed rotations along with the barriers to the rigid rotations in the clusters. Results of the four clusters follow a similar pattern. The barriers are lowered slightly by 1~2 kcal/mol in the first optimization step. There is a further decrease in the barriers around 7~10 kcal/mol in the second optimization step. The structural parameters provided by the X-ray diffraction experiments are confirmed as the ground state structure of the clusters. The decrease of the barriers is mainly caused by the relaxation of the transition states. For example, in the site 1 cluster of **1,2,4-OTMB**, there are three neighboring *t*-butyl groups reorienting slightly in the first optimization step. The nearest neighboring *t*-butyl group (in terms of the distance between the Ct carbon atoms of the groups, inter-Ct distance, 5.36 Å, in this case) reorients by about -5° in the transition state. The next nearest *t*-butyl group (inter-Ct distance, 5.96 Å) reorients by -2° and one of the third nearest *t*-butyl groups (inter-Ct distance, 6.27 Å) reorients by 1°. The change of the orientations of the hydroxyl groups and other structural parameters is negligible. In the second optimization step, two of the constituent methyl groups on the central *t*-butyl group show simultaneous reorientations, besides the reorientations of the *t*-butyl groups shown in the first step. As the central *t*-butyl group reaches its rotational transition state, the nearly in-plane methyl group (which becomes one of the out-of-plane methyl groups in the transition state) reorients by -4°, and one of the out-of-plane methyl groups (which becomes the in-plane methyl group in the transition state) reorients by 7°. Additionally, the rotation axis, the Ca-Ct bond, elongates slightly by 0.02 Å and opens its angle, α(Ct-C2-C1), by 6°. The latter change is presumably to relieve the repulsion between the in-plane methyl group and the oxygen atom of the hydroxyl group. Such a relaxation mechanism has been observed in the calculations of the isolated molecules (see the previous chapter), but there is only a small degree of decrease of the barriers (less than 1 kcal/mol) due to the relaxation. In contrast, in the clusters, the intramolecular relaxation relieves the barriers by 7~10 kcal/mol. We may argue that the intramolecular relaxation relieves not only the intramolecular repulsion but also the intermolecular repulsion as well, and it results in a significant decrease of the barriers in the clusters due to the relief of intermolecular interactions.

Structural relaxation in the other three clusters follows a similar pattern, though there is difference in some details. The optimization results can be found in Table 5.8.

Table 5.7. The barriers to the partially relaxed rotation of the central *t*-butyl groups in the clusters, computed in a two-step scheme

Cluster	Model	Barriers, kcal/mol		
		Rigid	1st Step	2nd Step
1,2,4-OTMB				
Site 1	HF/3-21G	23.8	22.8	15.2
	B3LYP/6-31G(d)	21.0		12.9
Site 2	HF/3-21G	23.2	22.1	15.2
	B3LYP/6-31G(d)	19.4		12.5
1,2,4,6-OTMTB				
2-*t*-butyl group	HF/3-21G	25.7	23.8	14.2
	B3LYP/6-31G(d)	22.3		11.6
6-*t*-butyl group	HF/3-21G	20.7	18.7	11.9
	B3LYP/6-31G(d)	17.7		9.49

Table 5.8. Significant[a] structural relaxation results during the rotation of the central *t*-butyl group in the clusters

No	Remark[b]	Parameter	TS-GS[c]	No	Remark[b]	Parameter	TS-GS[c]
1,2,4-OTMB							
		Site 1				**Site 2**	
j	5.36 Å	τ(CmCtCaC1)	-6.4°	g	5.34 Å	τ(CmCtCaC1)	-5.3°
		τ(CmCtCaC1)	-5.6°			τ(CmCtCaC1)	-4.6°
		τ(CmCtCaC1)	-5.2°			τ(CmCtCaC1)	-4.5°
h	5.96 Å	τ(CmCtCaC1)	-2.4°	b	Out→out	τ(HCmCtCa)	10.2°
		τ(CmCtCaC1)	-3.3°			τ(HCmCtCa)	11.6°
		τ(CmCtCaC1)	-2.5°			τ(HCmCtCa)	10.5°
a	In→out	τ(HCmCtCa)	**-3.7°**		In→out	τ(HCmCtCa)	3.8°
		τ(HCmCtCa)	-4.6°			τ(HCmCtCa)	5.4°
		τ(HCmCtCa)	-5.2°			τ(HCmCtCa)	4.2°
	Out→in	τ(HCmCtCa)	7.2°		Out→in	τ(HCmCtCa)	8.0°
		τ(HCmCtCa)	7.0°			τ(HCmCtCa)	6.9°
		τ(HCmCtCa)	7.1°			τ(HCmCtCa)	7.4°
		α(CtCaC1)	6.1°			CtCa	0.028 Å
						α(CtCaC1)	5.1°
1,2,4,6-OTMTB							
		2-*t*-butyl group				**6-*t*-butyl group**	
6	5.562 Å	τ(CmCtCaC1)	-2.6°	14	5.39 Å	τ(CmCtCaC1)	-3.3°
		τ(CmCtCaC1)	-2.0°			τ(CmCtCaC1)	-3.5°
		τ(CmCtCaC1)	-2.2°			τ(CmCtCaC1)	-3.5°
		τ(HOC1Ca)	9.7°			τ(HOC1Ca)	-4.1°
8	5.513 Å	τ(CmCtCaC1)	4.0°	18	5.35 Å	τ(CmCtCaC1)	5.3°
		τ(CmCtCaC1)	3.3°			τ(CmCtCaC1)	4.2°

No	Remark[b]	Parameter	TS-GS[c]	No	Remark[b]	Parameter	TS-GS[c]
		τ(CmCtCaC1)	3.8°			τ(CmCtCaC1)	4.5°
		τ(HOC1Ca)	3.6°			τ(HOC1Ca)	-4.5°
1		τ(HOC1Ca)	9.7°	19	5.53 Å	τ(CmCtCaC1)	3.8°
	Out→out	τ(HCmCtCa)	7.8°			τ(CmCtCaC1)	2.8°
		τ(HCmCtCa)	7.5°			τ(CmCtCaC1)	3.1°
		τ(HCmCtCa)	7.5°			τ(HOC1Ca)	-2.5°
	Out→in	τ(HCmCtCa)	4.4°	21	5.56 Å	τ(CmCtCaC1)	-2.0°
		τ(HCmCtCa)	1.4°			τ(CmCtCaC1)	-1.6°
		τ(HCmCtCa)	1.1°			τ(CmCtCaC1)	-1.8°
	In→out	τ(HCmCtCa)	-4.5°	2	Out→in	τ(HCmCtCa)	-10.2°
		τ(HCmCtCa)	-3.9°			τ(HCmCtCa)	-12.0°
		τ(HCmCtCa)	-3.2°			τ(HCmCtCa)	-9.8°
		CtCa	0.018 Å		In→out	τ(HCmCtCa)	-7.4°
		α(CtCaC1)	6.7°			τ(HCmCtCa)	-7.4°
						τ(HCmCtCa)	-8.2°
						CtCa	0.028 Å
						α(CtCaC1)	-5.6°

[a] Only the bond angle and dihedral changes greater than 2° and bond length changes greater than 0.01 Å are reported.

[b] The lengths refer to the inter-Ct distance of the *t*-butyl groups to the central *t*-butyl group. Otherwise, the groups are in the central *t*-butyl group or the central molecule. The comments, 'out→in', 'in→out' and 'out→out', refer that how the constituent methyl groups switch their positions between the in-plane and out-of-plane positions from the ground state to the transition state.

[c] The values are the difference between the transition state and the ground state in their respective partially optimized clusters.

5.5. DISCUSSION

5.5.1. Barriers and Intermolecular Interactions

The barriers to the rigid rotation of the *t*-butyl groups, the constituent methyl groups and the hydroxyl groups in the four clusters are collected in Table 5.6. Comparing the values with the barriers in the isolated molecules computed in the previous chapter, we find that intermolecular interactions dominate the barriers to the internal rotation of the overall *t*-butyl groups and hydroxyl groups. We may further decompose the barriers into the contribution from intramolecular interactions and the contribution from intermolecular interactions, or the contribution from the intermolecular interaction with each neighbor. This decomposition can be achieved through calculating the rigid rotation barriers in the reduced-size clusters by removing the neighbors selectively, as was done in Chapter 3. We may also do an approximate analysis to investigate how the interactions between the neighboring *t*-butyl groups affect their barriers by assuming a Lennard-Jones type potential. In the first-order approximation, we can evaluate the strength of the interactions between two *t*-butyl groups by the inter-Ct distance. This index successfully predicts how the neighboring *t*-butyl groups respond to the rotation of the central *t*-butyl group in the partially relaxed calculations of the clusters.

A more detailed and accurate analysis needs to consider the orientation of the *t*-butyl groups. By treating the constituent methyl groups as spherical objects centered at their carbon positions, we can draw the dependence of the inter-Cm distance on the orientation of the *t*-butyl groups. If the inter-methyl interactions can be modeled by a Lennard-Jones potential, and can be summed together as the estimation of the interaction between *t*-butyl groups, such inter-Cm distance graphs are helpful in evaluating the interaction between *t*-butyl groups, especially in the repulsion range, as shown in the examples given in Chapter 3. However, one should be cautious that there are also strong intermolecular interactions between the *t*-butyl groups and the benzene rings that could involve the rotation of the *t*-butyl groups in the solids of **1,2,4-OTMB** and **1,2,4,6-OTMTB**, as discussed in the case of 1,4-di-*t*-butylbenzene in Chapter 3. Through the analyses discussed above, we find that it is the collective result of the intermolecular interactions from several neighbors that determines the barrier to the rotation of the *t*-butyl groups. (For example, there are three neighboring *t*-butyl groups within the inter-Ct distance of 6.0 Å from the central *t*-butyl group in the site 1 cluster of **1,2,4-OTMB**. See Table 5.9. The value of 6.0 Å is a rough estimate of the diameter of a *t*-butyl group. In the site 2 cluster of **1,2,4-OTMB**, there are two inter-Ct distances shorter than 6.0 Å. For the 2-*t*-butyl group in **1,2,4,6-OTMTB**, there are four such closest neighboring *t*-butyl groups, and for the 6-*t*-butyl group, there are six such neighbors.) There is no dominant interaction from a single neighbor, although some of the neighbors in the first interaction shell have a greater contribution than others. This lack of a dominant factor leads to weak coupling between the internal rotation of the neighboring *t*-butyl groups, as shown in the partially relaxed calculations.

The internal rotation of the constituent methyl groups is only slightly affected by the intermolecular interactions. This result is not surprising since the size of the methyl groups is small and the intermolecular interactions between the methyl groups and their neighbors are not strongly dependent on the orientation of the methyl groups. This result has also been found in the other three aromatic compounds in Chapters 2 and 3.

Table 5.9. Six shortest inter-Ct distances between the neighboring *t*-butyl groups and the central *t*-butyl group in the clusters. The labels of the neighboring *t*-butyl groups are corresponding to those in A, B, C and D in Figure 5.1, respectively

1,2,4-OTMB (site 1)						
Neighbor	**j**	**m**	**h**	**c**	**d**	**i**
Inter-Ct distance, Å	5.338	5.756	5.976	6.262	6.262	6.566
1,2,4-OTMB (site 2)						
Neighbor	**g**	**k**	**j**	**n**	**h**	**d**
Inter-Ct distance, Å	5.338	5.976	6.262	6.262	6.566	6.959
1,2,4,6-OTMTB (2-*t*-butyl)						
Neighbor	**6**	**8**	**10**	**22**	**13**	**15**
Inter-Ct distance, Å	5.559	5.515	5.939	5.829	6.444	6.444
1,2,4,6-OTMTB (6-*t*-butyl)						
Neighbor	**14**	**18**	**19**	**21**	**13**	**9**
Inter-Ct distance, Å	5.372	5.372	5.515	5.559	5.829	5.939

5.5.2. Comparison between 1,2,4-OTMB and 1,2,4,6-OTMTB

In the isolated molecules, the symmetrical two *t*-butyl groups in **1,2,4,6-OTMTB** cause a crowded environment for the hydroxyl group and bring down the rotation barrier, compared with **1,2,4-OTMB**. In return, the low hydroxyl rotation barrier also affects the PES of the *t*-butyl groups and the methyl groups. In the crystals of the compounds, there are significant differences in the barriers (see Table 5.6) and the PESs between the two compounds, as shown in the Results section.

Although the densities of the two phenols are close (the density of **1,2,4-OTMB** is 1.04 times higher than that of **1,2,4,6-OTMTB**, as computed from the unit cell parameters in Table 5.1), the mean number of *t*-butyl groups per volume ('*t*-butyl density') in **1,2,4,6-OTMTB** is 1.43 times higher than that in **1,2,4-OTMB**. This relatively higher *t*-butyl density is because there are two *t*-butyl groups per molecule of **1,2,4,6-OTMTB**. The higher density means that the *t*-butyl groups pack in a more compact fashion in **1,2,4,6-OTMTB**. This point is also confirmed by the closest inter-Ct distances for the four types of *t*-butyl groups in the two compounds, as seen in Table 5.9. Especially, for the 6-*t*-butyl group in **1,2,4,6-OTMTB**, there are 6 close neighbors with the inter-Ct distance shorter than 6.0 Å to the central 6-*t*-butyl group. The crowded environment of this group again significantly lowers the barrier in the cluster, compared with the other three clusters. This result is not surprising as was discussed earlier in this thesis. Although the crowded environment may bring strong repulsions, the collective result depends on both the strength of the repulsions and their spatial distribution. Compact packing is more likely to distribute the repulsion forces evenly to achieve a more stable ground state conformation, and thus reduce their effect on the barrier. This contemplation is supported by the computed results of the partially relaxed rotation in the clusters. (See Table 5.8) In the 6-*t*-butyl group cluster, the four nearest *t*-butyl groups in term of the inter-Ct distance reorient by 2~5°. They are distributed roughly on the symmetric positions of the central *t*-butyl group. (That is, they are connected by the symmetry plane containing the bonds of Cm-Ct and Ct-Ca. The Cm refers to the in-plane methyl carbon.) In addition, two of the four nearest *t*-butyl groups are 6-*t*-butyl groups and the other two are 2-*t*-butyl groups. In the other three clusters, there are only one or two neighbor *t*-butyl groups reorienting in the central *t*-butyl rotation process, and the reorienting groups belong to the other type.

Intermolecular interactions exert strong effects on the internal rotation of the hydroxyl group in both the compounds and cause a significant increase of the barriers. The barrier in **1,2,4,6-OTMTB** is approximately half of that in **1,2,4-OTMB**. However, the two-fold symmetry in the isolated molecule of **1,2,4,6-OTMTB** is lost due to the unsymmetrical intermolecular interactions. The low barrier in **1,2,4,6-OTMTB** provides the possibility of coupling the hydroxyl group rotation with the *t*-butyl group rotation in the clusters as in the isolated molecules. As shown in Table 5.8, the hydroxyl groups on the nearest neighbors reorient by 3~10° as the rotation of the central *t*-butyl group.

5.5.3. Rotational Couplings in the Clusters

In the isolated molecules, the coupled rotation route between the *t*-butyl group and the constituent methyl groups is not a favorable one in both the phenols. The reason is that the barrier to the *t*-butyl group rotation increases to higher than 3.0 kcal/mol because of the ortho hydroxyl group. In the clusters, the computed *t*-butyl rotation barriers are nearly tripled because of the intermolecular interactions, while the barriers to the methyl rotations are almost as same as in the isolated molecules. As discussed in the previous chapter, the two-dimensional PES with respect to the internal rotation of the *t*-butyl group and the methyl groups can be viewed as a superimposed landscape of the constrained methyl rotation, which is dependent on the relative orientation the *t*-butyl group (τ(*t*-butyl)), on the rotational curve of the *t*-butyl group. The increase of the *t*-butyl rotation barrier does not change the overall shape of the PES but makes the coupled rotation route less likely. However, on the other hand, the small degrees of reorientation of the methyl groups are found to play an important role in lowering the barriers to the *t*-butyl rotation in the clusters. This coupled methyl orientation relaxation relieves the hindrance from the intramolecular interactions and from the intermolecular interactions as well, as shown by the results of the partially relaxed rotation. This coupling scheme is found to be common for the *t*-butyl group rotation on aromatic systems.

Besides the intramolecular coupling scheme, there is also the coupled orientation relaxation for the nearest *t*-butyl neighbors in the clusters. Although there is no dominant strong intermolecular interaction which can drive the neighboring *t*-butyl group to rotate cooperatively with the central *t*-butyl group, a few nearest *t*-butyl neighbors respond to the central *t*-butyl group rotation by reorienting 2~5°. This across-molecular relaxation scheme is more significant in **1,2,4,6-OTMTB** since the *t*-butyl groups pack more compactly.

In the isolated molecules of **1,2,4,6-OTMTB**, the hydroxyl group rotation barrier is small, and the coupled route between the rotation of the hydroxyl group and the rotation of the 2-*t*-butyl group is energetically preferred. The situation changes in the crystal solid. The preferred hydroxyl orientation is the nearly perpendicular conformation. Its relative positions to the two *t*-butyl groups are similar. The barrier also increases significantly due to the intermolecular interactions, but it is not so high as in **1,2,4-OTMB**. The partially relaxed rotation calculation shows that the hydroxyl group reorients by 10° as the 2-*t*-butyl group rotates in the cluster. For the rotation of the 6-*t*-butyl group in the cluster, the change of the hydroxyl group orientation is negligible. Additionally, the hydroxyl groups on the nearest neighbors reorient by 2 to 10° during the rotation of both the 2-*t*-butyl and 6-*t*-butyl groups. This across-molecular relaxation scheme could be a result of the intermolecular interactions between the hydroxyl groups and the central *t*-butyl group. It could be also driven by the intramolecular interactions between the hydroxyl groups and the *t*-butyl groups which reorient too as the central *t*-butyl group rotates. The hydroxyl group orientation relaxation, either the intramolecular relaxation or the across-molecular relaxation, is not observed in the clusters of **1,2,4-OTMB**, probably due to the very high barrier for the hydroxyl group.

5.6. CONCLUSION

Results in this chapter show that two phenols have a much higher barrier to the internal rotation of the *t*-butyl groups in their crystalline solids than the aromatic hydrocarbons of 1,4-di-*t*-butylbenzene, 2,6-di-*t*-butylnaphthalene and 3-*t*-butylchrysene. There is a large contribution from intermolecular interactions besides the contribution from intramolecular interactions. The rotation of the *t*-butyl groups is accompanied by a slight degree of across-molecular coupling of the *t*-butyl rotations and by the geared reorientations of the constituent methyl groups and the hydroxyl groups along other structural relaxations. The rotation of the hydroxyl groups is also strongly hindered by intermolecular interactions while the rotation of the constituent methyl groups is mainly governed by intramolecular interactions.

5.6. REFERENCES

[1] Beckmann, PA; Ratcliffe, CI; Dunell, B. Proton spin-lattice relaxation and methyl group rotation. *J. Magn. Reson.*, 1978, 32, 391-402.

[2] Beckmann, PA; Fusco, FA; O'Neill, AE. Proton spin-lattice relaxation and intramolecular reorientation in solids. *J. Magn. Reson.*, 1984, 59(1), 63-70.

[3] Beckmann, PA; Paty, C; Allocco, E; Herd, M; Kuranz, C; Rheingold, AL. The relationship between crystal structure and methyl and *t*-butyl group dynamics in van der Waals organic solids. *J. Chem. Phys.*, 2004, 120(11), 5309-5314.

[4] Beckmann, PA. Proton spin relaxation and methyl and hydroxyl group motion. *J. Magn. Reson.*, 1979, 36, 199-205.

[5] Beckmann, PA; Cheung, AM; Fisch, EE; Fusco, FA; Herzog, RE; Narasimhan, M. Methyl and *tert*-butyl reorientation and distributions of activation energies in molecular solids. A nuclear spin-relaxation study in 2,4- and 2,5-di-*tert*-butyl-hydroxybenzene. *J. Chem. Phys.*, 1986, 84(4), 1959-1968.

[6] Frisch, MJ; Trucks, GW; Schlegel, HB; Scuseria, GE; Robb, MA; Cheeseman, JR; Montgomery, Jr., JA; Vreven, T; Kudin, KN; Burant, JC; Millam, JM; Iyengar, SS; Tomasi, J; Barone, V; Mennucci, B; Cossi, M; Scalmani, G; Rega, N; Petersson, GA; Nakatsuji, H; Hada, M; Ehara, M; Toyota, K; Fukuda, R; Hasegawa, J; Ishida, M; Nakajima, T; Honda, Y; Kitao, O; Nakai, H; Klene, M; Li, X; Knox, JE; Hratchian, HP; Cross, JB; Bakken, V; Adamo, C; Jaramillo, J; Gomperts, R; Stratmann, RE; Yazyev, O; Austin, AJ; Cammi, R; Pomelli, C; Ochterski, JW; Ayala, PY; Morokuma, K; Voth, GA; Salvador, P; Dannenberg, JJ; Zakrzewski, VG; Dapprich, S; Daniels, AD; Strain, MC; Farkas, O; Malick, DK; Rabuck, AD; Raghavachari, K; Foresman, JB; Ortiz, JV; Cui, Q; Baboul, AG; Clifford, S; Cioslowski, J; Stefanov, BB; Liu, G; Liashenko, A; Piskorz, P; Komaromi, I; Martin, RL; Fox, DJ; Keith, T; Al-Laham, MA; Peng, CY; Nanayakkara, A; Challacombe, M; Gill, PMW; Johnson, B; Chen, W; Wong, MW; Gonzalez, C; Pople, JA. *Gaussian 03, Revision B.05*, Gaussian, Inc., Pittsburgh PA, 2003.

[7] The X-ray crystal structures are available free of charge from the Cambridge Crystallographic Data Centre. See http://www.ccdc.cam.ac.uk/ for details.

[8] Rheingold, AL; Figueroa, JS; Dybowski, C; Beckmann, PA. Superlattices, polymorphs and solid-state NMR spin-lattice relaxation (T_1) measurements of 2,6-di-*tert*-butylnaphthalene. *Chem. Comm.*, 2000, (8), 651-652.

[9] Beckmann, PA; Burbank, KS; Clemo, KM; Slonaker, EN; Averill, K; Dybowski, C; Figueroa, JS; Glatfelter, A; Koch, S; Liable-Sands, LM; Rheingold, AL. ^{1}H nuclear magnetic resonance spin-lattice relaxation, ^{13}C magic-angle-spinning nuclear magnetic

resonance spectroscopy, differential scanning calorimetry, and X-ray diffraction of two polymorphs of 2,6-di-*tert*-butyl-naphthalene. *J. Chem. Phys.*, 2000, 113(5), 1958-1965.

[10] Iimura, Y; Sakurai, T; Ohno, Y; Asahi, K.-I; Isono, K. *Acta Crystallogr., C.*, 1983, 39, 778.

[11] Bolte, M; Amon, M. *Private Communication*, 2000.

[12] Maze-Baudet, M. *Acta Crystallogr., B.*, 1973, *29*, 602.

[13] Abragam, A. *The Principles of Nuclear Magnetism.* Oxford University Press: Oxford, 1961.

[14] Slichter, CP. *Principles of Magnetic Resonances.* 3rd ed.; Springer: Berlin, 1990.

[15] Beckmann, PA; Hill, AI; Kohler, EB; Yu, H. Nuclear-spin relaxation in molecular solids with reorienting methyl and *t*-butyl groups: The spectral density and the state of the solid. *Phys. Rev. B*, 1988, 36(16), 11098-11111.

[16] Beckmann, PA; Buser, CA; Gullifer, K; Mallory, FB; Mallory, CW; Rossi, GM; Rheingold, AL. Methyl and *t*-butyl group reorientation in planar aromatic solids: low-frequency nuclear magnetic resonance relaxometry and X-ray diffraction. *J. Chem. Phys.*, 2003, 118(24), 11129-11138.

Chapter 6

INTERNAL ROTATION OF THE CF_3 GROUP IN 3-(TRIFLUOROMETHYL)PHENANTHRENE

6.1. ABSTRACT

The internal rotation of the CF_3 group around the C3-Cm bond in 3-(trifluoromethyl)phenanthrene is studied by quantum mechanical electronic structure calculations. The rotational potential energy surfaces were calculated for the isolated molecule and a 13-molecule cluster model. The barrier in the isolated molecule is 0.66 kcal/mol at the HF/6-311+G(d,p)//HF/6-31G(d) level and 0.46 kcal/mol at the MP2/6-31G(d)//MP2/6-31G(d) level. The barrier in the cluster model is 2.62 kcal/mol, which agrees well with the experimental value, 2.8 ± 0.2 kcal/mol, obtained by nuclear spin relaxometry. The intermolecular interaction between the nearest pair of the CF_3 groups contributes approximately 75% of the barrier in the cluster. The nearest CF_3 pair is found to undergo cooperative reorientations. The disorder of the CF_3 groups found in the crystal is explained by the strong interaction between the closest pair of the molecules and the rotational dynamics of the groups.

6.2. INTRODUCTION

The focus of this chapter is on the internal rotation of the trifluoromethyl group (CF_3) in 3-(trifluoromethyl)phenanthrene (**3-F3P**) in both the isolated molecule and the crystalline solid. Compared with hydrogen, fluorine has a much larger electronegativity and a somewhat longer van der Waals radius. Comparison studies on the methyl and trifluoromethyl groups may provide more insights on the internal rotation dynamics.

The crystal structure of **3-F3P** was measured by single crystal X-ray diffraction, which yields the thermal equilibrium structure at the measurement temperature [1]. Because of the much higher electron density on fluorine than on hydrogen, the coordinates of the trifluoromethyl group are expected to be measured accurately, in contrast to the methyl group. However, there is a dynamical disorder problem involved which is common to the hindered rotor with a small to intermediate barrier [2]. There is a wide angular distribution for the orientation of the CF_3 groups in the crystal. (Though such a thermal distribution has the potential to be used to evaluate the rotational barrier, this was not done in this study.) The X-ray diffraction post-process software generated two conformations for the CF_3 group with a population ratio of 80/20.

In the crystal of **3-F3P**, the relaxation of the nuclear spins of F-19 and H-1 is mainly through the dipolar interactions between like spins (F-F and H-H) and unlike spins (F-H). Among the three types of the spin dipolar interactions, the interactions of F-F and F-H are modulated by the internal rotation of the CF_3 group in the crystal. By measuring the temperature dependence of the solid state F-19 and H-1 spin-lattice relaxation rate in the polycrystalline sample of the compound, the rotational barrier was determined. Through the measurements of both the F-19 and H-1 spin-lattice relaxation rates at two NMR frequencies (22.5 and 53.0 MHz) in the temperature range between 114 and 300 K, the barrier is found to be about 2.8 ± 0.2 kcal/mol [3].

Through *ab initio* electronic structure calculations, we wish to obtain the potential energy surface for the internal rotation of the CF_3 group and to partition the barrier into *intra*molecular and *inter*molecular interactions. In the crystals of methylphenanthrenes, the internal rotation of the methyl groups is governed largely by intramolecular interactions, with little involvement of intermolecular interactions [4]. Does this conclusion also hold true for the crystal of **3-F3P**, in which the CF_3 group has a bigger size? For the methyl group rotation in the isolated molecules, the hyperconjugation between the methyl group and the aromatic ring determines the ground state and the barrier for the rotation, in the absence of any steric effect. We also look at the role of the hyperconjugation interaction [5-8] in the internal rotation of the CF_3 group. For comparison, rotation barriers for CH_2F and CHF_2 groups were also calculated for the isolated molecules of 3-(fluoromethyl)phenanthrene (**3-F1P**) and 3-(difluoromethyl)phenanthrene (**3-F2P**) and also for the rotation barrier for the CF_3 group in 2-(trifluoromethyl)- phenanthrene. The results are compared with those calculated for 3-methyl-phenanthrene (**3-MP**).

6.3. Methods

Electronic structure calculations were carried out using *Gaussian03* software package [9] running under Redhat Linux OS on dual-processor Dell Precision 650 workstations. Initial geometries for the isolated molecule calculations were set up with the aid of *GaussView 3.0* (Gaussian, Inc., Wallingford, CT.). The Natural bond orbital (NBO) population analysis was processed with *NBO 3.1* [10] incorporated in *Gaussian03*. The cluster model of 3-(trifluoromethyl)phenanthrene for simulating the crystal packing environment was built from the X-ray crystal structure of the compound with home-written codes. The model included all the molecules with any atom falling into a sphere of radius 6.5 Å with its center located at the $C(sp^3)$ atom of the trifluoromethyl group. The radius was chosen to include a sufficient number, twelve in this case, of neighboring molecules to reasonably model the local environment for the CF_3 group of interest on the center molecule. The structural parameters of hydrogen atoms in the cluster were optimized at the HF/3-21G level since they are unavailable from the X-ray diffraction experiments.

Structural optimizations for the isolated molecules were carried out at several theoretical levels, HF/3-21G//HF/3-21G, HF/6-31G(d)//HF/6-31G(d), B3LYP/6-31G(d)// B3LYP/6-31G(d), B3LYP/6-311+G(d,p)// B3LYP/6-311+G(d,p), B97-2/6-311+G(d,p)//B97-2/6-

311+G(d,p) and MP2/6-31G(d)//MP2/6-31G(d). Calculations on cluster models were done using the HF/3-21G and B3LYP/6-31G(d) models. For the partially relaxed potential energy surface for clusters, optimization at the HF/3-21G//HF/3-21G level was carried out. The density functional theory, which is relatively more efficient and includes more electron correlation effects, was used to obtain barriers in the cluster models.

The potential energy surfaces were drawn with *Origin 7.0* (OriginLab Co., Northampton, MA.). The Renka-Cline gridding method was used to generate the two-dimensional potential energy surfaces. Other graphs were created with the aid of *Mathematica 5.0* (Wolfram Research, Inc., Champaign, IL.).

6.4. RESULTS

6.4.1. Computed and Experimentally Measured Molecular Structures

The ground state structure of the isolated molecule of **3-F3P** was obtained by a full optimization at the HF/6-31G(d)//HF/6-31G(d) level. The normal mode analysis at the HF/6-31G(d) level was used to confirm that the optimization produced the global minimum energy conformation. The parameters of the computed structures are compared with the X-ray diffraction determined structure in Table 6.1. The optimized structures at the HF/3-21G and B3LYP/6-311+G(d,p) levels are also listed for comparison. The numbering scheme of the molecule is shown below. For convenience, one of the dihedrals formed between the Cm-F bonds and the C3-C4 bond, τ(F-Cm-C3-C4), is used to label the orientation of the CF_3 group, τ(3-CF_3). The other two dihedrals are approximately equal to τ(3-CF_3) ± 120°, because the F-Cm-F bond angles are largely rigid when studying the internal rotation.

As seen in Table 6.1, the molecular structures for the isolated molecule computed at the three levels, including the HF/3-21G model, are in good agreement with the experimental structure as determined by X-ray crystallography except for the conformational orientation of the CF_3 group and some bond lengths. In the computed ground state conformation of the isolated molecule, one of the Cm-F bonds is coplanar and syn with the C3-C4 bond. In the crystalline solid state, none of the Cm-F bonds is coplanar with the phenanthrene ring as found by the X-ray diffraction method [1]. The situation in the crystal is complicated by the dynamic disorder problem of the CF_3 group. The X-ray data show two different conformations, one with a population of 80%, and the other with a population of 20%. In the 80% conformation, the dihedral of τ(F-Cm-C3-C4) was reported to be -33°, and in the 20% conformation, the dihedral was reported to be -48°. Another interpretation for the disorder of the CF_3 group would be that the actual conformation of the molecule in its lowest-energy thermal equilibrium orientation has a dihedral of τ(F-Cm-C3-C4) that lies between -33 and -48°, and the CF_3 group is undergoing conformational oscillations between these two limiting values, thereby resulting in disorder in the orientations of the CF_3 groups in the crystal as observed on the time scale of the X-ray photons.

To find the thermal equilibrium orientation of the CF_3 group in the crystal, the *ab initio* electronic structure computation was employed. First, a 13-molecule cluster model was built from the X-ray crystal structure of the compound. See Figure 5.1. The cluster, which includes all the 12 neighbor molecules in the first interaction shell of the central CF_3 group, is used to simulate the local environment for the CF_3 group in the crystal. Within this cluster model, the coordinates of all the CF_3 groups and hydrogen atoms were optimized at the HF/3-21G level independently, while keeping the carbon frame of phenanthrene rings fixed at their thermal equilibrium positions as found by the X-ray studies. Though the HF/3-21G model describes intermolecular interactions poorly, it produces molecular structures quite accurately. More importantly, it gives the same trend for the potential energy surface for the internal rotation in the cluster, compared with the B3LYP/6-31G(d) model, which includes more electron correlation effect.

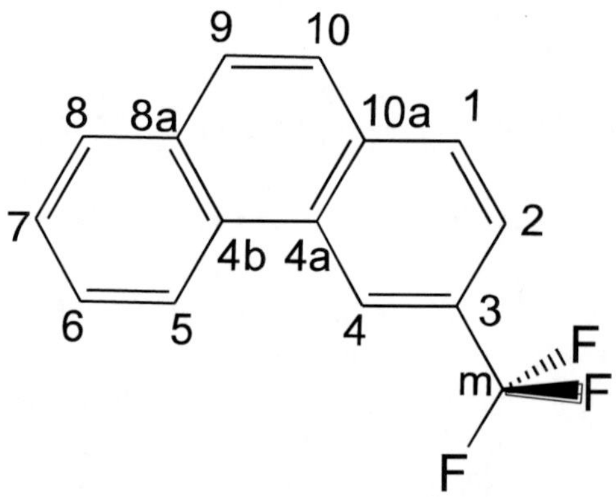

Scheme 6.1.

Table 6.1. Comparison between the calculated structures of the isolated molecule and the X-ray diffraction results for crystalline 3-(trifluoromethyl)phenanthrene

Parameter[a]	HF/ 3-21G	HF/ 6-31G(d)	B3LYP/ 6311+G(d,p)	X-ray[b]
Bond length, Å				
C1C2	1.360	1.361	1.375	1.364
C2C3	1.398	1.406	1.409	1.398
C3C4	1.361	1.365	1.380	1.375
C4C4a	1.407	1.412	1.412	1.403
C4aC4b	1.458	1.460	1.456	1.454
C4aC10a	1.402	1.402	1.423	1.418
C4bC5	1.408	1.411	1.413	1.403
C4bC8a	1.403	1.404	1.424	1.418
C5C6	1.366	1.367	1.381	1.376
C6C7	1.400	1.403	1.406	1.395
C7C8	1.364	1.365	1.378	1.359
C8C11	1.406	1.409	1.413	1.409
C11C9	1.438	1.440	1.433	1.427
C9C10	1.338	1.340	1.358	1.344
C10C10a	1.438	1.439	1.433	1.432
C10aC1	1.409	1.411	1.415	1.402
C3Cm	1.483	1.501	1.503	1.490

Parameter[a]	HF/ 3-21G	HF/ 6-31G(d)	B3LYP/ 6311+G(d,p)	X-ray[b]
CmF1	1.351	1.323	1.351	1.316(1.332)
CmF2	1.354	1.326	1.357	1.345(1.315)
CmF3	1.354	1.326	1.357	1.357(1.310)
Bond angle,°				
α(F1CmC3)	112.1	112.7	113.0	109.3(113.5)
α(F2CmC3)	112.0	111.5	111.8	116.0(111.8)
α(F3CmC3)	112.0	111.5	111.8	111.9(113.9)
Bond dihedral,°				
τ(F1CmC3C4)	0.0	0.0	0.0	87.7(70.4)
τ(F2CmC3C4)	-120.8	-120.6	-120.6	-153.5(-166.3)
τ(F3CmC3C4)	120.8	120.6	120.6	-32.6(-47.9)

[a] Numbering scheme as in the figure in text.
[b] Parenthetic values refer to the CF_3 orientation with the population of 20%.

The structural optimization results in a distribution of orientations among the 13 CF_3 groups. The computed conformation of the central CF_3 group, which has the most similar local environment of the CF_3 group in the crystal, represents the preferred orientation in the crystal. The optimized orientation of the central CF_3 group is with $\tau(CF_3)$ = -20°. For the other CF_3 groups in the cluster, the *absolute* value of the orientation dihedral ranges from 15 to 35°, which is the direct consequence of the local environments being different from the local environment of the central CF_3 group. Significantly, we found the dihedral of the closest neighboring CF_3 group is 28°.

6.4.2. Potential Energy Surfaces in the Isolated Molecules

The relaxed potential energy surfaces (PESs) for the CF_3 group in the isolated molecules of 3-(trifluoromethyl)phenanthrene and 2-(trifluoromethyl)-phenanthrene were computed by fixing one of the τ(F-Cm-C3-C4) dihedrals in 3-(trifluoromethyl)-phenanthrene or one of the τ(F-Cm-C2-C1) dihedrals in 2-(trifluoromethyl)phenanthrene while fully optimizing all the other structural parameters. Because of the local 3-fold symmetry of the CF_3 group, the results in the range from $\tau(CF_3)$ = 0 to 60° provide the full information of the potential energy surfaces. (The CF_3 group is effectively rigid for the rotation about the symmetry axis. The three-fold symmetry is further supported by the fitting results of the potential energy surfaces as shown later, in which only three-fold potential is needed.) The potential energy values in other ranges can be easily obtained by the symmetry property.

The PESs computed at the HF/6-311+G(d,p)//HF/6-31G(d) level for the internal rotation of the CF_3 group in the isolated molecules of the two compounds are shown in Figure 6.2. The two curves nearly overlap. The structures of the ground states and the transition states for **3-F3P** and **2-F3P** are shown in Scheme 6.2. The barrier to the internal rotation of the CF_3 groups, defined as the energy difference between the transition state and the ground state, is 0.66 kcal/mol in both **3-F3P** and **2-F3P**. The potential energy surfaces can be fitted to a three-fold cosine potential function plus a very small magnitude of a six-fold cosine potential function, $V(\tau)=\frac{V_3}{2}(1-\cos 3\tau)+\frac{V_6}{2}(1-\cos 6\tau)$. The best-fit results are drawn in Figure 6.2 in solid lines. The best-fit coefficients, V_3 and V_6, are 0.66 ± 0.001 and -0.04 ± 0.001 kcal/mol, respectively, in **3-F3P**, and 0.66 ± 0.001 and -0.06 ± 0.001 kcal/mol, respectively, in **2-F3P**.

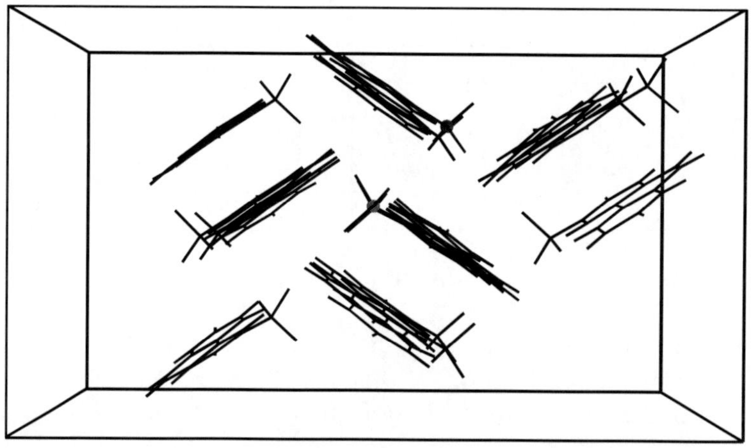

(A) 13-molecule cluster model of 3-(trifluoromethyl)phenanthrene

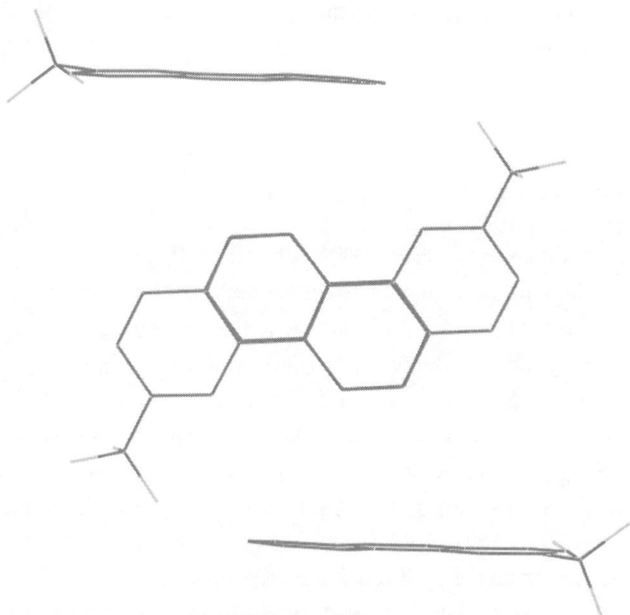

(B) Crystal packing pattern in a unit cell

Figure 6.1. Perspective projection of the 13-molecule cluster model (**A**) and parallel projection of a unit cell (**B**) in the crystal of 3-(trifluoromethyl)-phenanthrene. For clarity, hydrogen is not shown and fluorine is in green in **B**. In **A**, the central CF_3 group is labeled by a red dot and the closest neighbor CF_3 group labeled by a blue dot.

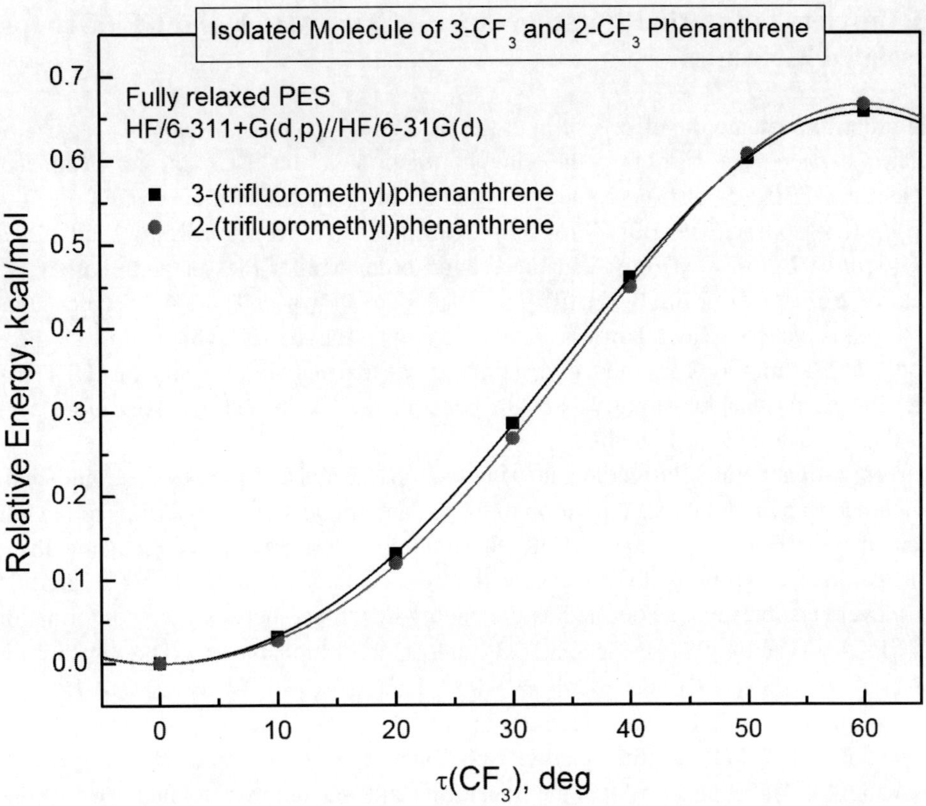

Figure 6.2. Fully relaxed potential energy surfaces for the internal rotation of 3-trifluoromethyl and 2-trifluoromethyl groups in the isolated molecules of 3-(trifluoromethyl)phenanthrene and 2-(triflu oromethyl)phenanthrene, computed at the HF/6-311+G(d,p)//HF/6-31G(d) level. The rotational dihedral $\tau(CF_3)$ refers to the dihedral of τ(F-Cm-C3-C4) in 3-(trifluoromethyl)phenanthrene and τ(F-Cm-C2-C1) in 2-(trifluoromethyl)phenanthrene, respectively. Solid lines are the least-square best fit results with the three-fold and six-fold cosine potential functions, $V(\tau)=\frac{V_3}{2}(1-\cos 3\tau)+\frac{V_6}{2}(1-\cos 6\tau)$.

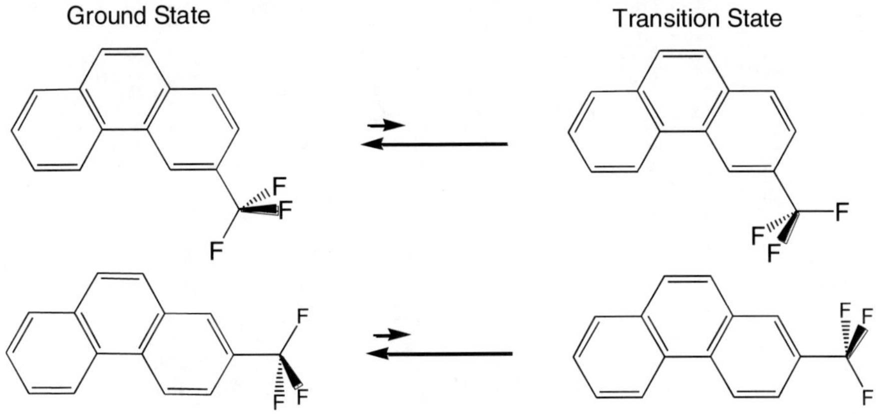

Scheme 6.2.

6.4.3. Barriers to Internal Rotation of CF_3, CHF_2, CH_2F and CH_3 Groups in the Isolated Molecules

In order to evaluate the effects of both the theoretical model and the basis set size on the calculated barriers, the barrier to the internal rotation of the 3-CF_3 group in an isolated molecule of **3-F3P** was also computed with the density functional theory (DFT), B3LYP/6-31G(d)//B3LYP/6-31G(d), B3LYP/6-311+ G(d,p)//B3LYP/6-311+G(d,p) and B97-2/6-311+G(d,p)//B97-2/6-311+G(d,p) and the second order Møller-Plesset perturbation method (MP2), MP2/6-31G(d)// MP2/6-31G(d). The results are shown in Table 6.2. The calculations at all levels give consistent barriers. Compared with the barrier obtained with the MP2 method, the Hartree-Fock method overestimates the barrier slightly and the DFT method underestimates the barrier slightly [11]. In general, the DFT method gives more accurate results than the Hartree-Fock method.

To investigate the intramolecular interactions between the CF_3 group and the phenanthrene ring on the internal rotation in the isolated molecules, the barriers to the internal rotation of 3-CHF_2, 3-CH_2F and 3-CH_3 phenanthrenes, systematically replacing the three fluorines of the CF_3 group by hydrogens, were also calculated at the HF/6-311+G(d,p)//HF/6-31G(d) level. The barriers, as defined as the energy difference between the conformations of $\tau(3\text{-}CH_nF_{3-n}) = 0$ and 180° (see Scheme 6.3), are reported in Table 6.3. The dihedral of $\tau(3\text{-}CH_nF_{3-n})$ refers to τ(H-Cm-C3-C4) for **3-F2P** and τ(F-Cm-C3-C4) for **3-F1P**. The 0° conformations of the three compounds with n = 1 to 3 are the ground state. The 180° conformation of **3-MP** is the methyl rotational transition state. Whether the 180° conformation of **3-F2P** and **3-F1P** is the transition state was not investigated. The differences among the barriers are quite small, and it appears that there is no simple dependence on the relative numbers of fluorine and hydrogen atoms. Further discussion on this can be found in the discussion section.

Table 6.2 Barrier (in kcal/mol) to the internal rotation of the trifluoromethyl group in the isolated molecule of 3-(trifluoromethyl)phenanthrene computed at a series of theoretical levels

Theoretical Method	Barrier, kcal/mol
HF/6-311+G(d,p)//HF/6-31G(d)	0.66
B3LYP/6-31G(d)// B3LYP/6-31G(d)	0.40
B3LYP/6-311+G(d,p)//B3LYP/6-311+G(d,p)	0.40
B97-2/6-311+G(d,p)// B97-2/6-311+G(d,p)	0.40
MP2/6-31G(d)// MP2/6-31G(d)	0.46

Table 6.3. Barrier to internal rotation of the methyl and fluorine substituted methyl groups in isolated molecules of phenanthrene derivative compounds, computed at the HF/6-311+G(d, p)//HF6-31G(d) theoretical level

Compound Name and Abbreviation	Barrier[a], kcal/mol
2-(trifluoromethyl)phenanthrene, **2-F3P**	0.66
3-(trifluoromethyl)phenanthrene, **3-F3P**	0.66
3-(difluoromethyl)phenanthrene, **3-F2P**	0.72

Compound Name and Abbreviation	Barrier[a], kcal/mol
3-(fluoromethyl)phenanthrene, **3-F1P**	0.75
3-methylpheanthrene, **3-MP**	0.54

[a] Energy difference between the ground state and the transition state or the 180° conformation.

Ground State

$\tau(3\text{-CH}_n\text{F}_{3\text{-}n}) = 0°$, $n = 1\sim3$ $\tau(3\text{-CH}_n\text{F}_{3\text{-}n}) = 180°$

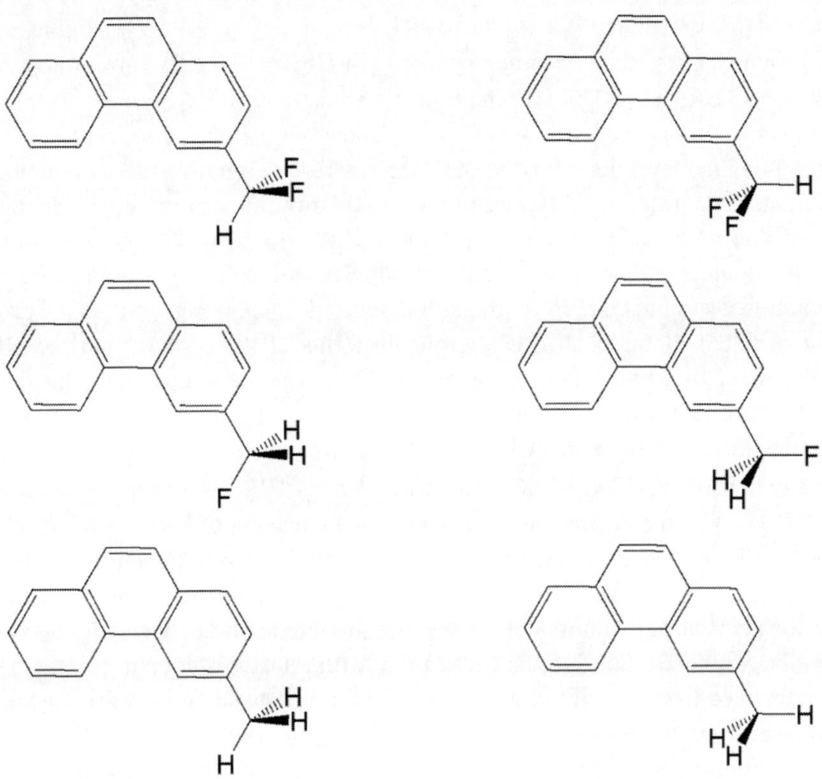

Scheme 6.3.

6.4.4. Barriers to Internal Rotation of CF_3 Group in the Clusters

The barrier to the internal rotation of the CF_3 group in the crystal of **3-F3P** was studied in the 13-molecule cluster. The cluster was built as stated earlier. The structural coordinates of the CF_3 groups were optimized at the HF/3-21G level in the cluster. Since the local environment of the central CF_3 group most closely matches the environment found in the crystal, the optimized structure of the central CF_3 group was chosen for all 12 of the neighboring molecules in the cluster. The underlying assumption is that all the CF_3 groups have the same thermal equilibrium orientation in the crystal. We will find later that this assumption does not apply to the crystal of **3-F3P**.

Two kinds of PESs, rigid and partially relaxed, were computed for the internal rotation of the central CF_3 group in the cluster. The rigid PES was calculated by stepwise reorientation of the central CF_3 group while keeping all the other parameters of the cluster fixed. To calculate the partially relaxed PES, the structural parameters of all 13 CF_3 groups, including the bond lengths, bond angles and dihedrals, were allowed to be optimized at the HF/3-21G level during the manipulated rotation of the central CF_3 group.

In Figure 6.3, the PESs obtained at the HF/3-21G and B3LYP/6-31G(d) levels are plotted for both the rigid and partially relaxed rotations. For the rigid rotation, the computed barrier is 8.52 kcal/mol at the B3LYP/6-31G(d) level. (At the HF/3-21G level the barrier is slightly smaller. However, the rigid PESs at the two levels of theory follow a similar trend.) The rotational ground state is the conformation with $\tau(3\text{-}CF_3) = -20°$. By turning the central CF_3 group around its C3-Cm bond by 50°, the transition state with $\tau(3\text{-}CF_3) = 30°$ is reached.

When all CF_3 groups were allowed to be optimized in the cluster, the barrier computed at the B3LYP/6-31G(d) level decreased to 2.62 kcal/mol. The optimized CF_3 orientations on the 12 neighbors in the cluster are different by 10 to 20° from the ground state orientation in the crystal, as would be expected from the possible edge effects in this small cluster. There is only one CF_3 group that changes its orientation significantly when the central CF_3 group goes from its rotational ground state to its transition state. It is identified as the CF_3 group that is the closest neighbor of the central CF_3 group in terms of the distance between the carbon atoms of the two CF_3 groups. This distance is 4.07 Å. The value is about 1.4 Å shorter than the distance to the next-nearest neighbor CF_3 group, which is 5.45 Å. At the rotational ground state of the central CF_3 group, $\tau(3\text{-}CF_3) = -20°$, the optimization finds the nearest CF_3 group has an orientation of $\tau(3\text{-}CF_3) = 28°$. When the central CF_3 group reaches to its transition state, $\tau(3\text{-}CF_3) = 30°$, the optimized orientation of the nearest CF_3 group is $\tau(3\text{-}CF_3) = 45°$, which is via a 17° reorientation from its ground state orientation in the cluster. It is concluded the reorientation of the nearest CF_3 group removes most of the 8.52 kcal/mol barrier that was calculated for the rigid rotation model. Therefore, the interaction between the two CF_3 groups may have a largest contribution to the barrier. To further test this conclusion, the nearest CF_3 group was removed from the cluster and replaced by a hydrogen atom at the 3-position of the phenanthrene ring. In the absence of the nearest CF_3 group, the rigid rotation barrier for the central CF_3 group falls to 2.50 kcal/mol at the B3LYP/6-31G(d) level, which is a decrease of 6.02 kcal/mol from the barrier in the original cluster. The barriers for the different calculation schemes are collected in Table 6.4.

Table 6.4 Barrier to the internal rotation of the trifluoromethyl group on the central molecule in the 13-molecule cluster model of 3-(trifluoromethyl) phenanthrene, computed at the B3LYP/6-31G(d) level. See the text for detailed descriptions of the computation schemes

Computation Scheme	Barrier, kcal/mol
Rigid rotation	8.52
Partially relaxed rotation at the HF/3-21G level	2.62
Rigid rotation, after removing nearest CF_3 group	2.50

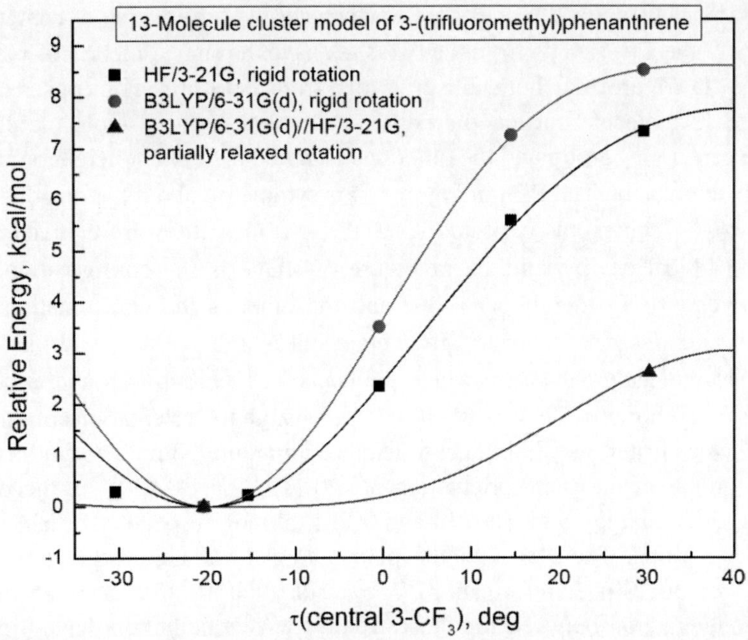

Figure 6.3. Potential energy surfaces for the internal rotation of the central CF_3 group in the 13-molecule cluster model of 3-(trifluoromethyl)phenanthrene, computed at the HF/3-21G and B3LYP/6-31G(d) levels. For the rigid rotation, all the structural parameters are fixed in the cluster when rotating the central CF_3 group. For the partially relaxed rotation, all the CF_3 groups are optimized at the HF/3-21G level at the ground state and the transition state for the central CF_3 group rotation. See the text for details. The solid lines are only to serve as guides for the data points.

6.4.5. Two-Dimensional PES in the Closest-Pair Cluster

As shown by both the rigid rotation calculation and the partially relaxed rotation calculation in the previous section, the intermolecular interaction between the central CF_3 group and its closest neighboring CF_3 group in the cluster is the dominant factor that determines the equilibrium orientation of the central CF_3 group and its rotational barrier. A two-molecule cluster, which is composed with the closest pair of molecules in the crystal, was then built to model such an interaction. See Scheme 6.4. The PES with respect to the rigid rotations of the two CF_3 groups for every 10° was calculated at the B3LYP/6-31G(d) level and it is shown in Figure 6.4.

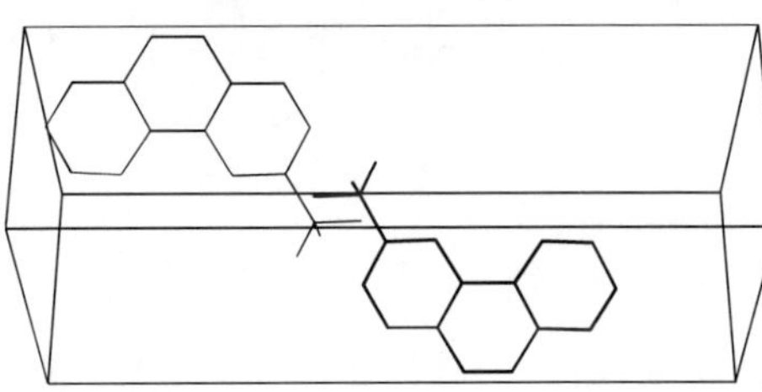

Scheme 6.4.

The two molecules of this pair are connected by an inversion center. The same conformations of the CF_3 groups from the two sites have the same dihedral of $\tau(CF_3)$ but with opposite signs. In Figure 6.4, both dihedrals run from 0 to 60°. For consistency with the labeling in the 13-molecule cluster, the negative dihedral in the 2-molecule cluster is also labeled as the center CF_3 group and the other one is called the closest neighbor, although they are completely equivalent. The PES in Figure 6.4 is symmetric about the diagonal line, $\tau(CF_3$, neighbor) = $-\tau(CF_3$, center), as expected since the two molecules are equivalent. There are two equivalent global energy minima in Figure 6.4. One is the conformation with $\tau(CF_3$, center) = -20° and $\tau(CF_3$, neighbor) = 30° and the other is the conformation with $\tau(CF_3$, center) = -30° and $\tau(CF_3$, neighbor) = 20°. This result is in agreement with the optimization results in the 13-molecule cluster, in which the central CF_3 group is found at $\tau(CF_3)$ = -20° and the closest neighbor at $\tau(CF_3)$ = 28°. If we interchange the orientations of the closest pair in the 13-molecule cluster, we should get to another minimum point. The conformations with both the CF_3 groups at the same orientation of -20 (20)° or -30 (30)° in the two-molecule cluster are higher in energy by about 0.13 and 0.12 kcal/mol, respectively, than the energy of the two global minima. There is a saddle point along the diagonal direction close to the conformation of -30 (30)° conformation. The results tell us that because of the strong intermolecular interaction between the closest pair, the two neighbors have a preference to take slightly different orientations with 10° apart. This is probably the energetic reason for the disorder of the CF_3 group in the crystal found in the X-ray data. The conformation with both CF_3 groups at 0°, which is the ground state conformation for the isolated molecule, is the global energy maximum point with a relative energy 10.1 kcal/mol higher than the global minimum.

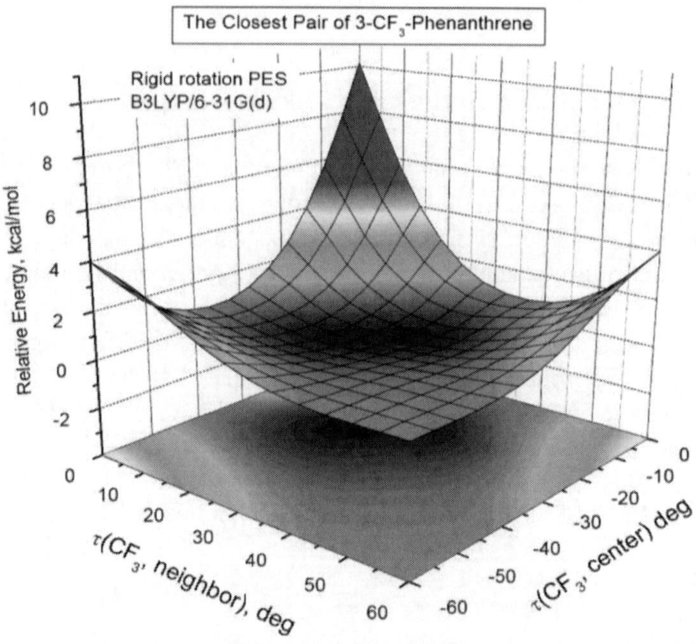

Figure 6.4. The two-dimensional potential energy surface of the closest pair cluster in Scheme 6.4 with respect to the rigid rotation dihedrals for every 10° calculated at the B3LYP/6-31G(d) level.

The two-dimensional PES in Figure 6.4 not only exposes the energetic reason for the two CF_3 orientations in the thermal equilibrium structure of the crystal, the 'disorder problem', but also indicates that the rotations of the CF_3 groups on the closest pair are likely coupled. In the partially relaxed rotation of the central CF_3 group, the closest neighbor CF_3 group was found to reorient by 15° at the rotational transition state of the central CF_3 group. In Figure 6.4, it is clear that the minimum energy path involves the two global energy minima, which can interconvert easily bypassing a saddle point of about 0.12 kcal/mol height. The coupling between the two groups alleviates the possible strong repulsion between them.

6.5. Discussion

6.5.1. Role of Hyperconjugation in Internal Rotation in Isolated Molecules

The internal rotation of the methyl group on aromatic systems has been studied extensively. The hyperconjugation between the methyl group and the aromatic ring is the main factor determining the barrier [12-18]. It is found that the barrier in absence of a steric effect is proportional to the bond order difference between the two aromatic ring C-C bonds adjacent to the methyl group [15]. In the isolated molecules, the barrier is in the range from 0.1 to 1 kcal/mol. If the two bonds are nearly equivalent, as in toluene, the barrier is negligible. (Both experimental measurements and theoretical studies have found that the barrier in toluene is only a few calories per mol. At room temperature, the methyl group in toluene behaves like a free rotor.)

Compared with hydrogen, fluorine has a much greater electronegativity. The C-H bond usually has a very small bond dipole, but in a C-F bond the fluorine atom has a significant fraction of a negative charge. It also costs 25 kcal/mol more energy to break the C-F bond than the C-H bond [19-21]. The van der Waals radius of fluorine is 1.47 Å, about 0.27 Å larger than hydrogen (1.20 Å) [22-23]. Substituting hydrogen atoms with fluorine atoms will alter the interaction between the methyl group and the aromatic ring [24-25]. We expect different barrier heights for the CH_3, CH_2F, CHF_2 and CF_3 groups. As seen in Table 6.3, the barrier for the trifluoromethyl group (0.66 kcal/mol) is slightly higher than for the methyl group (0.54 kcal/mol). The partially substituted fluoromethyl and difluoromethyl groups each have a higher barrier than the trifluoromethyl group, 0.75 kcal/mol for the fluoromethyl group and 0.72 kcal/mol for the difluoromethyl group. The differences are small.

The barriers could be understood by a more detailed analysis of the interaction components between the rotor group and the aromatic ring. Particularly, we focus on the steric effect and the hyperconjugative interaction. In Table 6.5, the nonbonded distances between the hydrogen atom on C4 or C2 of the phenanthrene ring and the three different atoms (H or F) on the rotor group on C3 of the ring are listed for both the rotational ground and transition states for the rotation. In 3-methylphenathrene, the shortest nonbonded H···H distance is about 2.30 Å in both conformations, which is close to the equilibrium van der Waals distance, 2.40 Å. Furthermore, there is no large difference in the nonbonded distances between the ground and transition states. We expect that there is no obvious steric effect involved in the internal rotation of the methyl group. For the fluoromethyl group, though the nonbonded inter-atom distances are similar as for the methyl group, some intramolecular repulsion exists between the ring H and the rotor F, because the equilibrium van der Waals distance is about 2.67 Å (1.20 Å + 1.47 Å). The transition state is less destabilized by the repulsive interaction since the calculated nonbonded H···F distance is longer by 0.05 Å. The same situation also occurs for the 3-fluoromethyl and 3-(difluoromethyl)phenanthrene as well. The steric effects analyzed from the nonbonded distances predict the ground state (0° conformation) is less stable than the 180° conformation. But the steric effects are then counteracted by the hyperconjugation between the rotor group and the phenanthrene ring.

Two of the C-H or C-F bonds in the methyl or fluorine-substituted methyl groups are symmetric about the phenanthrene ring plane in the ground states (0° conformation) and the transition states (180° conformation). The occupied σ orbitals of the C-H or C-F bonds and the lone pair orbitals on fluorine atoms with π symmetry could mix together and interact with the unoccupied π^* orbitals on the phenanthrene ring, provided the energies of the orbitals are close and there is significant overlapping. As a consequence, the electron density gets redistributed among the ring and the group. The electron re-distribution can also happen between the unoccupied σ^* orbitals of the C-H or C-F bonds with π symmetry and the occupied π orbitals on the phenanthrene ring. The hyperconjugation is a stabilization factor for the molecule. The magnitude of the hyperconjugation is estimated by second order perturbation theory in the Natural Bond Orbital (NBO) analysis [26-34]. The second order perturbation energy is given by,

$$\Delta E^{(2)} = -2\frac{<1|\hat{F}|2^*>^2}{\varepsilon_{2^*} - \varepsilon_1},$$

where $\hat{F}$ is the Fock operator and ε_1 and ε_{2^*} are NBO orbital energies for the occupied orbital |1> and the unoccupied orbital |2*>. The absolute values of the hyperconjugation energy between the rotor group and the phenanthrene ring are tabulated in Table 6.6. The calculations were done at the HF/6-311+G(d,p) level.

As seen in Table 6.6, hyperconjugation for **3-MP** exists between the occupied σ C-H orbitals of the methyl group and the unoccupied π* orbital of the ring for the methyl group. The interaction is stronger at the ground state by 1.22 kcal/mol than the transition state. For the trifluoromethyl group, the electron redistribution direction is opposite, from the occupied π orbital on the ring to the unoccupied σ^* orbitals of the C-F bonds, due to the electronegativity of fluorine. There is also a slight electron back donation from the lone electron pair on the fluorine and the occupied σ orbital to the π^* orbital of the ring. As for the methyl group, the strongest hyperconjugation also occurs at the ground state, for which the perturbation energy is higher by 3.29 kcal/mol than the transition state. The hyperconjugations in the fluoromethyl and difluoromethyl groups are similar as in the trifluoromethyl group though the contributions of each component are different. The difluoromethyl group is more like the trifluoromethyl group. In the fluoromethyl group, there are two C-H bonds. The hyperconjugation between them and the ring are similar as found for the methyl group. In addition, there is also a significant interaction between the π orbitals of the ring and the σ^* orbitals of the C-F bonds. The hyperconjugation explains why the barriers for the fluorine substituted methyl groups are similar to the barriers for the methyl group even though fluorine is very different from hydrogen chemically.

6.5.2. Disorder of the CF_3 Group in Crystalline Solids

The results section shows that there is an energetic reason, two global energy minima, for the disorder of CF_3 groups in the crystal. However, there is also a well known thermodynamic disorder related to the rotation motion. For the hindered internal rotation with a low to intermediate barrier, there is a wide distribution of the molecules over the orientation dihedral angles of the rotor. In Figure 6.5, the calculated conformational distribution of the CF_3 group at 298 K is plotted for three barrier heights, 0.46, 3.0 and 6.0 kcal/mol. A three-fold rotational potential energy is used and the phenanthrene ring of the molecule is assumed to be rigid. The orientation is in a Gaussian-type distribution around the bottom of the three potential wells. At the barrier height of 3.0 kcal/mol, the width at half-height of the distribution peaks is about 28°. The distribution at 213 K, the measurement temperature of the X-ray diffraction experiment, is slightly narrower but the main feature remains same. The barrier to the internal rotation of the CF_3 group in the crystal has been found experimentally by NMR relaxation measurements to be about 2.8 ± 0.2 kcal/mol, which is consistent with the computed barrier of 2.62 kcal/mol in the cluster model. Therefore, there exists a significant probability for the rotational orientation of the CF_3 group to be found within ±15° from the thermal equilibrium conformation.

The calculations in the 13-molecule cluster and the 2-molecule closest pair show there are two preferred orientations in the clusters, $\tau(CF_3)$ = -20° or -30°. The X-ray diffraction pattern of the crystal was fitted with two orientations, $\tau(CF_3)$ = -33° and -48°. The former is with 80% population and the latter with 20%. The difference between the former conformation and the computed orientation is about 13°, which is within the dynamics disorder range of the CF_3 group.

Table 6.5. Smallest non-bonded distances between each of the three H or F atoms in a CH_nF_{3-n} (n = 0~3) group and two ring hydrogen atoms in the rotational ground and transition states or 180° conformation of the isolated molecules. The geometries were optimized at the HF/6-31G(d) level. See Schemes 6.2 and 6.3 for the structures

Distance, Å	3-MP	3-F1P	3-F2P	3-F3P	2-F3P
Ground State	2.346(H···H)	2.330(H···F)	2.334(H···H)	2.346(H···F)	2.387(H···F)
	2.770(H···H)	2.745(H···H)	2.801(H···F)	2.795(H···F)	2.801(H···F)
	2.770(H···H)	2.745(H···H)	2.801(H···F)	2.795(H···F)	2.801(H···F)
Transition State[a]	2.381(H···H)	2.364(H···F)	2.389(H···H)	2.393(H···F)	2.396(H···F)
	2.733(H···H)	2.708(H···H)	2.755(H···F)	2.750(H···F)	2.799(H···F)
	2.733(H···H)	2.708(H···H)	2.755(H···F)	2.750(H···F)	2.786(H···F)

[a] The transition state or the 180° conformation.

Table 6.6. Hyperconjugative interactions between the rotor group and the phenanthrene rings, calculated by the second order perturbation theory with the Natural Bond Orbital (NBO) analysis at the HF/6-311+G(d,p)//HF/6-31G(d) level

Compound		Second Order Perturbation Energy, kcal/mol	
		Ground State	Transition State[a]
3-MP	σ→π*	8.98	7.76
3-F1P	σ→π*	8.96	7.47
	lp→π*		
	π→σ*	7.30	5.45
3-F2P	σ→π*	2.06	1.54
	lp→π*	1.54	
	π→σ*	12.84	11.38
3-F3P	σ→π*	1.71	1.24
	lp→π*	1.82	
	π→σ*	12.82	11.82

[a] The transition state or the 180° conformation.

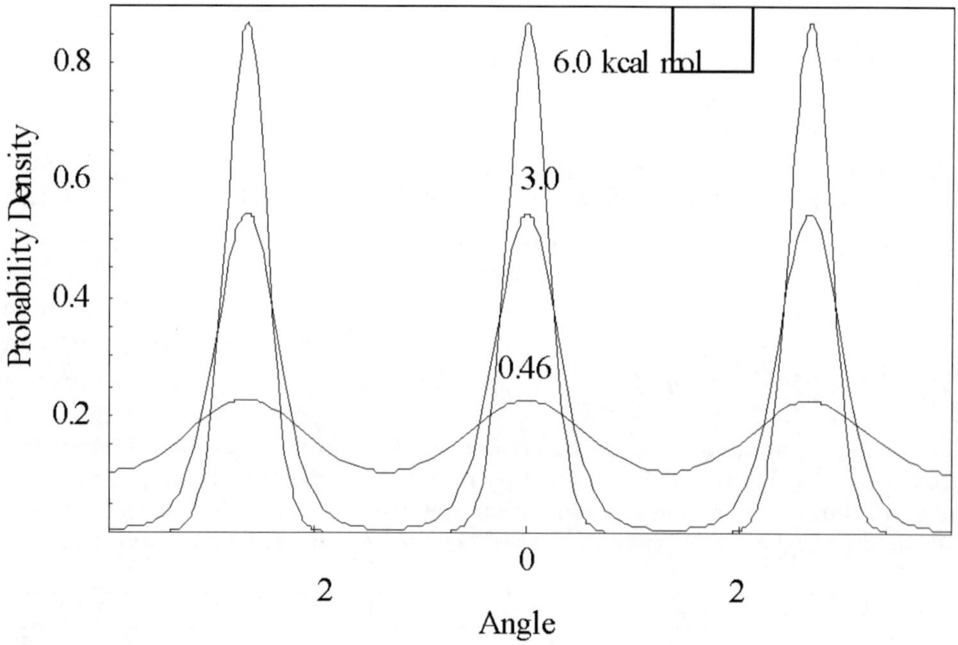

Figure 6.5. Probability density distribution of the orientation of the trifluoromethyl group at 298 K with the three-fold potential energy, $V_0(1\text{-}\cos 3\tau)/2$. The barriers V_0 are shown in the graph.

6.5.3. Intermolecular Interactions Involved in CF_3 Group Rotation in Crystalline Solids

The crystal structure of **3-F3P** was determined by the single crystal X-ray diffraction method at 213 K [1]. It was found that the crystal belongs to the monoclinic space group of

$P2_1/c$, with the following unit cell parameters, a = 11.1557(8), b = 12.5428(9) and c = 8.0282(6) Å, α = 90, β = 97.4500(10), and γ = 90°, V = 1113.85 Å^3 and Z = 4.

There is one unique molecular site in the crystal. All CF_3 groups share a common local environment except those on the crystal surfaces. There are four molecules in a unit cell. They pack in a pattern shown in Figure 6.1**B**. The phenanthrene rings of the four molecules form four surface planes of a hexahedron. The other planes of the hexahedron are open, with two CF_3 group sitting on each of the planes. In the crystal, the hexahedron structure is repeated in the three crystal directions. We will see there are 'walls' of the phenanthrene hexahedron. Between the two neighbor walls, there are two parallel layers of the CF_3 group from the face-to-face sides of the walls. The two layers are very close spatially and form a pseudo two-dimensional network of the CF_3 groups. In Figure 6.6, a projection picture of the two layers is made on a plane that divides the two layers equally. Only seven of the thirteen CF_3 groups in the cluster are shown in Figure 6.6. The other six groups are not in pseudo 2-dimensional plane. For clarity, only fluorine atoms are shown, which are represented by circles with radii given by the van der Waals radius of 1.47 Å of fluorine. Each triplet of circles represents a CF_3 group. The green CF_3 groups are from the molecules with the phenanthrene rings beneath the plane, and blue CF_3 groups are from the molecules with the phenanthrene rings above the plane. The intensity of the color is proportional to the absolute distance of the fluorine atom from the plane. The signs on the disk refers to the fluorine above (+) or below (-) the plane. The rotational axis of the CF_3 group, the C3-Cm bond, is not perpendicular to the plane, but has an angle about 45° for all CF_3 groups shown. For each CF_3 group, there are two fluorine atoms with approximately the same distance to the plane, but on the two sides of the plane. They are differentiated by the sign. The light blue or green color stands for a distance of 0.17 Å to the plane and the darker colors stand for 1.05 and 1.26 Å. Particularly, we notice there is a pair of CF_3 groups that are very close to each other in Figure 6.6 at least as projected on this plane. One is green and the other blue. If we index the neighbors with the inter-carbon distance between the CF_3 groups, which is invariant under CF_3 rotation, the distance is 4.08 Å for the pair of the closest neighbors. It is shorter by 1.37 Å than the distance between the next closest neighbors, which is 5.45 Å. The closest CF_3 pair happens to have the shortest distances between the fluorine atoms as well, as shown later. In the 13-molecule cluster, there is another pair of the closest neighbors besides the central CF_3 group with its nearest neighbor. These two pairs are circled out with the dotted ellipsoid in Figure 6.6.

A view of the inter-fluorine distances provides more information on the intermolecular interaction between CF_3 groups. This information is presented in Figure 6.7, which shows the inter-fluorine distances dependent on the rotational angles of the neighbor CF_3 groups from their computed equilibrium orientation, $\tau(3\text{-}CF_3)$ = -20°. Only the four closest pairs of neighbors are shown in Figure 6.7. Others are too far away and would have no steric effect on the rotation of the central CF_3 group. The corresponding inter-carbon distance for each neighbor is shown on the top of each graph. Each graph is composed of 9-subgraphs, which corresponds to the 9 pairs of inter-fluorine distances between the two CF3 groups. The distance is represented by color, as shown by the legend spectrum. The van der Waals equilibrium distance between two fluorine atoms is about 2.94 Å, which is on the turning edge from green to light blue in the spectrum of colors in Figure 6.7. For distances shorter than the equilibrium distance, there exists a strong repulsive interaction which is indicated by the blue to red color in the graph. The distance in the green-to-yellow color involves the weak long-range attractive interaction. For the yellow color, the distance is too large, and the associate intermolecular interaction is negligible. It is seen in Figure 6.7 that the blue-to-red region, the strong repulsion hot spot, only exists between the closest neighbor pairs. This leads the expectation that the interaction between the central CF_3 group and the nearest neighbor is the main contributor to the barrier to the central CF_3 group rotation in the cluster. This point is supported by the computational results.

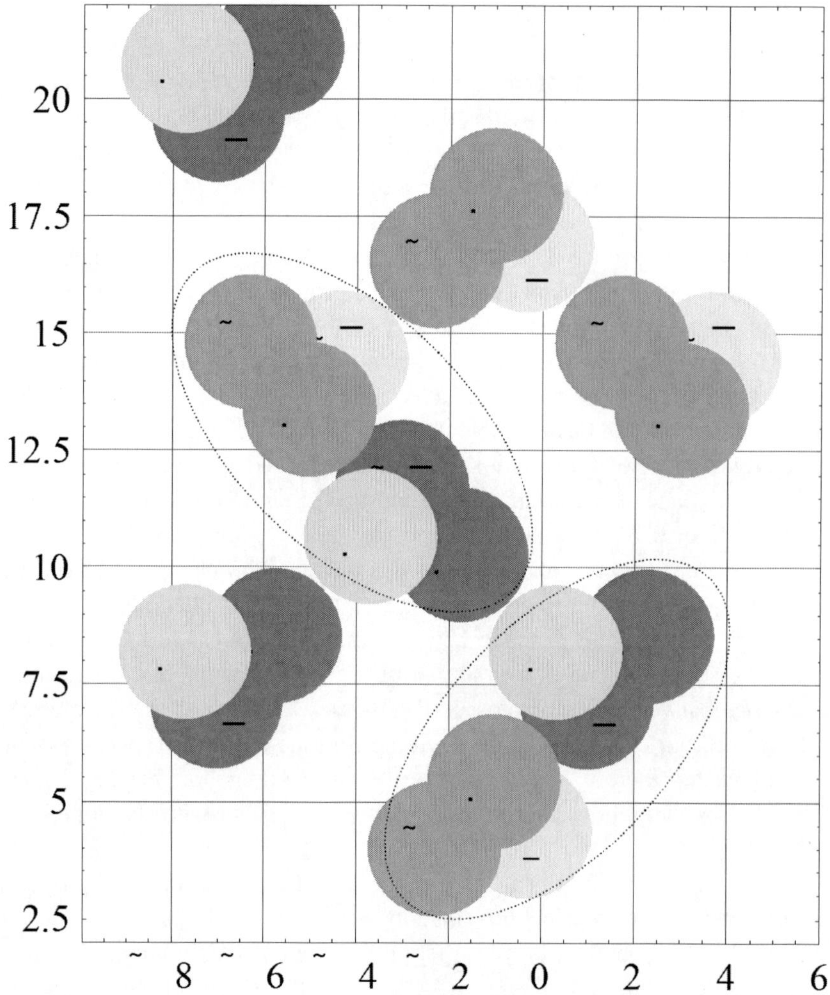

Figure 6.6. Projection of the two close-packed layers CF_3 groups in the cluster of 3-(trifluoromethyl)phenanthrene. Only fluorine atoms are shown. They are shown to scale the van der Waals radius. See text for details. The pairs of the nearest neighbors are circled by the dot ellipsoid. Legend, in Å.

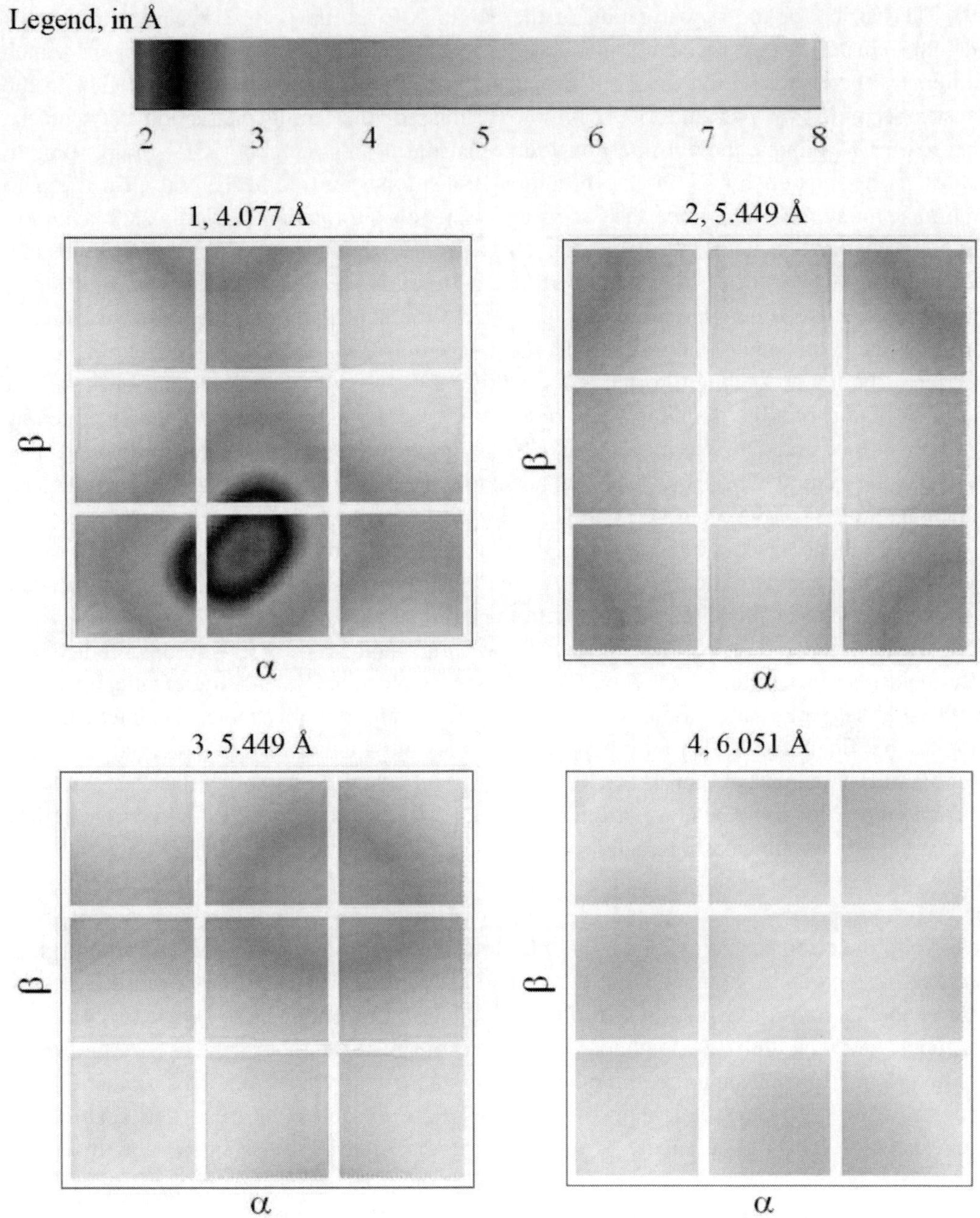

Figure 6.7. Inter-fluorine distance dependence of the rotational angles between the CF_3 group to the four nearest neighbors in the crystal of 3-(trifluoromethyl)-phenanthrene. The inter-carbon distances between the CF_3 groups are shown on the top of each graph. The angle α stands for the rotated angle (0 to 120° for each sub-graph) of the central CF_3 group from its computed equilibrium orientation, and the angle β for the neighbor. The distance is represented by color.

The barrier for the rigid rotation of the central CF_3 group is 8.52 kcal/mol in the 13-molcule cluster. After replacing the closest neighbor CF_3 group with hydrogen, which eliminates the repulsive interaction between the two, the barrier to the rigid rotation in the cluster decreases by 70% to 2.50 kcal/mol. Because of the strong interaction between the these two CF_3 groups, the optimization found that the nearest neighbor CF_3 group needs to adjust its orientation by 25° at the rotational transition state of the central CF_3 group to minimize the repulsion between the two groups. The relaxed rotation barrier is 2.62 kcal/mol, in good agreement with the value of 2.8 kcal/mol obtained experimentally by NMR relaxometry. The barrier may be partitioned into *intra*molecular and *inter*molecular interactions. By assuming the intramolecular interaction contribution in the cluster is same as in the isolated molecule, 0.66 kcal/mol, the intermolecular interaction contribution to the barrier is about 1.96 kcal/mol, which is about 75% of the total barrier. This value is consistent with the value of 70% found by the rigid rotation calculation scheme. The interactions between the central CF_3 group and other 11 neighbor CF_3 groups in the cluster and also between the central CF_3 group and the phenanthrene rings have little involvement in the rotational process of the central CF_3 group in the cluster.

In the above discussion, we have seen that the structural relaxation of the nearest neighbor CF_3 group is critical to relieve the repulsive interaction. This results in a calculated barrier in the cluster model that agrees with the solid state NMR result. This point is further supported by the intermolecular interaction potential analysis. Several groups studied the intermolecular interaction in CF_4 dimers and other molecular systems systematically using Møller-Plesset perturbation theory [35-37]. They found that the intermolecular interaction is able to be fitted to the sum of two interactions, the Lennard-Jones potential plus the electrostatic potential between F and F, C and C, and F and C. Although the CF_3 group on phenanthrene ring could be quite different from CF_4, we may use this CF_4 model potential to estimate the intermolecular interaction between the CF_3 groups. By applying this model and using the NBO charge on the isolated molecule of **3-F3P**, we found that the interaction between the central CF_3 group and the neighbor CF_3 groups is about -0.45 kcal/mol (it is relative to the reference zero value at an infinite distance and the minus sign says the groups are stabilized by their interactions) at the ground state in the cluster. If the central CF_3 group rotates to the transition state in a rigid fashion, the interaction energy increases to 2.56 kcal/mol. Among the 12 neighbors, only the nearest neighbor makes a significant contribution. The contributions from other neighbors are nearly zero. When using the relaxed orientation for the nearest neighbor, the interaction energy decreases from 2.56 kcal/mol to 0.62 kcal/mol. This rough model again says that the intermolecular interaction from the nearest CF_3 group neighbor involves is important in the rotation of the central CF_3 group in the crystal, and the pairs of CF_3 groups reorient in a concerted fashion.

6.6. CONCLUSION

In this chapter, the potential energy surface for the internal rotation of the CF_3 group around its C3-Cm bond in **3-F3P** was calculated in the isolated molecule and the 13-molecule cluster model. The following points could be concluded from this comparison study.

(1) The rotational barrier for the isolated molecule is 0.66 kcal/mol obtained at the HF/6-311+G(d,p)//HF/6-31G(d) level. The barrier is from the intramolecular interaction between the CF_3 group and the phenanthrene ring. The ground state is the conformation with one of the Cm-F bonds coplanar and syn with respect to the C3-C4 bond. The rotational transition state is reached by rotating the CF_3 group around the C3-Cm bond by 60°, which puts one of the Cm-F bonds coplanar and syn with respect to the C3-C2 bond. The potential energy surface has 3-fold symmetry.

(2) The 13-molecule cluster and the 2-molecule closest pair are used to understand CF_3 rotation in the crystal. It is found that due to the strong interaction between CF_3 groups with their closest neighboring CF_3 group, there are two preferred orientations of equal energy in the crystal, one with τ(F-Cm-C3-C4) = −20 and the other τ(F-Cm-C3-C4) = −30°. The two conformations have equal, 50%, population in the crystal.

(3) The rigid rotational barrier to the CF_3 rotation in the cluster is 8.50 kcal/mol, as computed at the B3LYP/6-31G(d) level. The partially relaxed rotational barrier is 2.62 kcal/mol, which is in good agreement with the value of 2.73 ± 0.1 kcal/mol measured by the NMR relaxation method.

(4) Intermolecular interactions contribute about 75% to the barrier of the CF_3 group in the crystal. Furthermore, the intermolecular interaction is mainly from the repulsion between the closest pair of the CF_3 groups in the crystal. The closest pair has an inter-carbon distance of 4.08 Å. The interactions from other neighbors have no significant contribution to the barrier.

(5) The closest pairs of the CF_3 group are found to rotate in a concerted fashion, because of the strong repulsive interaction between the two groups. The thermal equilibrium orientations of a closest pair are different by 10°. One is $\tau(CF_3)$ = −20° and the other one is $\tau(CF_3)$ = 28°. At the rotational transition state for the central CF_3 group in the cluster, the nearest CF_3 group adjusts its orientation by 17° from its equilibrium orientation. The pair is ubiquitous in the crystal and the interaction between the two attributes to the double minima for the CF_3 group rotation in the crystal and hence the disorder of the groups.

6.7. REFERENCES

[1] Wang, X; Mallory, FB; Mallory, CW; Beckmann, PA; Rheingold, AL; Francl, MM. CF_3 rotation in 3-trifluoromethylphenanthrene: X-ray diffraction and ab initio electronic structure. *J. Phys. Chem. A*, 2006, 110(11), 3954-3960.

[2] Trueblood, KN; Dunitz, JD. Internal molecular motions in crystals. The estimation of force constants, frequencies and barriers from diffraction data. A feasibility study. *Acta Crystallographica B*, 1983, B39(1), 120-133.

[3] Beckmann, PA; Rosenberg, J; Nordstrom, K; Mallory, CW; Mallory, FB. CF_3 rotation in 3-CF_3-phenanthrene: Solid state F-19 and H-1 NMR relaxation and Bloch-Wangsness-Redfield theory. *J. Phys. Chem. A*, 2006, 110(11), 3947-3953.

[4] (a) Conn, KG; Beckmann, PA; Mallory, CW; Mallory, FB. Methyl reorientation in methylphenanthrenes. I. Solid state proton spin-lattice relaxation in the 3-methyl, 9-methyl and 3,9-dimethyl systems. *J. Chem. Phys.* 1987, *87*(1), 20-27. (b)

Computational results (unpublished data, Wang, X. et al. 2005) from the isolated molecules and the cluster models of 3-methyl, 9-methyl and 3,9-dimethyl phenanthrene compounds show that for the rotations of either 3- or 9-methyl group in the crystals, the contribution of the intermolecular interactions is about 0.5 kcal/mol or less.

[5] Hoffmann, R; Pople, JA; Radom, L; Schleyer, PV; Hehre, WJ; Salem, L. Strong conformational consequences of hyperconjugation. *J. Am. Chem. Soc.*, 1972, 94(17), 6221-6222.

[6] Pophristic, V; Goodman, L. Hyperconjugation not steric repulsion leads to the staggered structure of ethane. *Nature*, 2001, 411, 565-568.

[7] Sadlej-Sosnowska, N. Energy barriers to internal rotation: Hyperconjugation and electrostatic description. *J. Phys. Chem. A*, 2003, 107, 8671-8676.

[8] Goodman, L; Pophristic, V; Weinhold, F. Origin of methyl internal rotation barriers. *Acc. Chem. Res.* 1999, 32(12), 983-993.

[9] Frisch, MJ; Trucks, GW; Schlegel, HB; Scuseria, GE; Robb, MA; Cheeseman, JR; Montgomery, Jr., JA; Vreven, T; Kudin, KN; Burant, JC; Millam, JM; Iyengar, SS; Tomasi, J; Barone, V; Mennucci, B; Cossi, M; Scalmani, G; Rega, N; Petersson, GA; Nakatsuji, H; Hada, M; Ehara, M; Toyota, K; Fukuda, R; Hasegawa, J; Ishida, M; Nakajima, T; Honda, Y; Kitao, O; Nakai, H; Klene, M; Li, X; Knox, JE; Hratchian, HP; Cross, JB; Bakken, V; Adamo, C; Jaramillo, J; Gomperts, R; Stratmann, RE; Yazyev, O; Austin, AJ; Cammi, R; Pomelli, C; Ochterski, JW; Ayala, PY; Morokuma, K; Voth, GA; Salvador, P; Dannenberg, JJ; Zakrzewski, VG; Dapprich, S; Daniels, AD; Strain, MC; Farkas, O; Malick, DK; Rabuck, AD; Raghavachari, K; Foresman, JB; Ortiz, JV; Cui, Q; Baboul, AG; Clifford, S; Cioslowski, J; Stefanov, BB; Liu, G; Liashenko, A; Piskorz, P; Komaromi, I; Martin, RL; Fox, DJ; Keith, T; Al-Laham, MA; Peng, CY; Nanayakkara, A; Challacombe, M; Gill, P; MW; Johnson, B; Chen, W; Wong, MW; Gonzalez, C; Pople, JA. *Gaussian 03, Revision B.05* Gaussian, Inc., Pittsburgh PA, 2003.

[10] Glendening, ED; Reed, AE; Carpenter, JE. Weinhold F. *NBO Version 3.1.*

[11] Zhao, Y; Pu, J; Lynch, BJ; Truhlar, DG. Tests of second-generation and third-generation density functionals for thermochemical kinetics. *Phys. Chem. Chem. Phys.*, 2004, 6, 673-676.

[12] Ito, M. in *Dynamics of Excited Molecules*, edited by Kuchitsu, K. Elsevier Science, Holland, 1994, Chap. 4, 7-120.

[13] Spangler LH; Pratt, DW. in *Jet Spectroscopy and Molecular Dynamics*, edited by Hollas, J. M.; Phillips, D., Blackie Academic and Professional, Glasgow, 1995, 369-398.

[14] Liljefors, T; Allinger, NL. *J. Comput. Chem.*, 1985, 6, 478.

[15] George, P; Bock, CW; Stezowski, JJ; Hildenbrand, T; Glusker, JP. *J. Phys. Chem.* 1988, 92, 5656.

[16] Lu, KT; Weinhold, F; Weisshaar, JC. *J. Chem. Phys.* 1995, 102, 6787.

[17] Nakai, H; Kawai, M. *Chem. Phys. Lett.*, 1999, 307, 272.

[18] Nakai, H; Kawai, M. *J. Chem. Phys.* 2000, 113, 2168.

[19] Hiyama, T. *Organofluorine Compounds: Chemistry and Applications*. Springer-Verlag: New York, 2000.

[20] Kraft, BM; Jones, WD. Mechanism of Vinylic and Allylic Carbon-Fluorine Bond Activation of Non-Perfluorinated Olefins Using $Cp^*{}_2ZrH_2$. *J. Am. Chem. Soc.* 2002, 124, 8681-8689.

[21] Farnham, WB; Smart, BE; Middleton, WJ; Calabrese, JC; Dixon, DA. *J. Am. Chem. Soc.*, 1985, 107, 4565.

[22] Bondi, A. *J. Phys. Chem.*, 1964, 68, 441-452.

[23] Rowland RS; Taylor, R. *J. Phys. Chem.*, 1996, 100, 7384-7391.

[24] Buono, RA; Zauhar, RJ; Venanzi, CA. *Ab initio* rotational barriers and solvation free energies of fluorinated dimethyl ethers. *J. Mol. Struct. (Theochem)*, 1996, 370, 97-133.

[25] Ignacio, EW; Schlegel, HB. *Ab initio* study of the structures, properties, and heats of formation of fluorochloromethanes and -silanes, $CH_{4\text{-}m\text{-}n}F_mCl_n$, and $SiH_{4\text{-}m\text{-}n}F_mCl_n$, (m + n = 0-4). *J. Phys. Chem.*, 1992, 96, 5830-5837.

[26] Reed, AE; Weinhold, F. *J. Chem. Phys.*, 1983, 78, 4066.

[27] Reed, AE; Weinhold, F. *J. Chem. Phys.* 1985, 83, 1736.

[28] Reed, AE; Weinstock, RB; Weinhold, F. *J. Chem. Phys.*, 1985, 83, 735.

[29] Reed, AE; Curtiss, LA; Weinhold, F. *Chem. Rev.*, 1988, 88, 899.

[30] Weinhold, F; Carpenter, JE. *Plenum*, 1988, 227.

[31] Carpenter, JE; Weinhold, F. *J. Mol. Struct.* (*Theochem*) 1988, 169, 41.

[32] Carpenter, JE. *PhD thesis*, University of Wisconsin, Madison, WI, 1987.

[33] Foster, JP; Weinhold, F. *J. Am. Chem. Soc.*, 1980, 102, 7211.

[34] The X-ray crystal structures are available free of charge from the Cambridge Crystallographic Data Centre. See http://www.ccdc.cam.ac.uk/ for details. The Cambridge Crystallographic Data Centre, Cambridge, UK.

[35] Parra, RD; Zeng, XC. Basis set effects on the intermolecular interaction of the CF_4–CF_4 system. *J. Mol. Struct. (Theochem)*, 2000, 503, 213-220.

[36] Tsuzuki, S; Uchimaru, T; Mikami, M; Urata, S. *Ab initio* calculations of intermolecular interaction of CHF_3 dimer: Origin of attraction and magnitude of CH/F Interaction. *J. Phys. Chem. A*, 2003, 107, 7962-7968.

[37] Urata, S; Tsuzuki, S; Mikami, M; Takada, A; Uchimaru, T; Sekiya, A. Analysis of the intermolecular interaction between CH_3OCH_3, CF_3OCH_3, CF_3OCF_3, and CH_4: High level *ab initio* calculations. *J. Comput. Chem.*, 2002, 23, 1472-1479.

Appendix

QUANTUM MECHANICS OF INTERNAL ROTATION

A.1. INTRODUCTION

The dynamics of the hindered methyl group and other rotor groups has been studied extensively and reviewed several times [1-6]. In this Appendix, the methodology is briefly reviewed and extended to the treatment of the internal rotation of the trifluoromethyl and *t*-butyl groups studied in this work. There are four degrees of internal rotational freedom in a *t*-butyl group. The overall internal rotation of the *t*-butyl group and individual rotation of the constituent methyl groups are first discussed separately as one-dimensional rotors. To evaluate the dynamical coupling between the *t*-butyl group and the constituent methyl groups, the two-dimensional rotor model needs to be considered [7-9].

Within the Born-Oppenheimer approximation, the motion of nuclei of a molecular system is separated from that of the electrons. The nuclear Hamiltonian energy operator is then written as the sum of the nuclear kinetic energy operator and the potential energy operator, $\hat{H} = \hat{T} + \hat{V}$. The potential energy can be obtained by solving the electronic structure of the molecular system with the theoretical methods discussed in the text. In Chapters 2-6, the potential energy surfaces for the internal rotation of the *t*-butyl, methyl and trifluoromethyl groups on a series of compounds were calculated in the isolated molecules and the cluster models, which are used to model the rotation in the gas phase and the crystalline state, respectively. Only the ground state electronic structure is discussed there, which is relevant in the nuclear magnetic resonance relaxation experiments. For some molecular spectroscopy methods, *e.g.* supersonic jet laser spectroscopy and fluorescence spectroscopy [10-16], the internal rotational dynamics also involves the transition between the electronic ground state and the excited states, which is beyond the scope of this study.

Focusing on the internal rotation dynamics, we may isolate the motion from other vibrational degrees of freedom of the molecular system. The equilibrium structure is used, assuming that only the ground state is populated for other vibrational modes. The coupling relationship between the internal rotation and the other vibrational modes is discussed in the context of the relaxed potential energy surfaces of the molecular systems in Chapters 2-6. When studying the thermodynamics of the internal rotation, the vibrational normal modes in the gas phase or the phonon spectrum in the solid phase may be treated as a lattice thermometer model without resorting to the fine structure of the vibration modes [17-18]. The coupling between the internal rotation and the rest vibration modes of the system maintains the thermal equilibrium.

In this Appendix, only the spatial wave functions of the internal rotors are discussed. A discussion of the spin wave functions and the coupling between the spin dynamics and the rotor dynamics, which leads to the spin relaxation, can be found in the literature [19-21].

A.2. Solution of the Schrödinger Equation for the One-Dimensional Internal Rotor

As a 0th-order approximation, the Schrödinger equation of the internal rotation of a methyl group or the overall rotation of a *t*-butyl group can be written as,

$$\hat{H}\Phi = E\Phi$$
$$\hat{H} = -\frac{\hbar^2}{2I}\frac{d^2}{d\tau^2} + V \tag{1}$$

where I is the reduced moment of inertia about the rotational axis parameterized by the angle τ. The wave function $\Phi(\tau)$ should satisfy the following natural periodic boundary condition,

$$\Phi(\tau) = \Phi(\tau + 2k\pi), \tag{2}$$

where k is integer. In the absence of the hindering potential function, V, Eq. 1 reduces to the free rotor problem. The general solution for the wave function of a free-rotor is,

$$\Phi_m(\tau) = \frac{1}{\sqrt{2\pi}} e^{im\tau}, \quad m = 0, \pm 1, \pm 2, \ldots \tag{3}$$

where m is the rotational quantum number. The eigenvalues of the energy are,

$$E_m = \frac{m^2\hbar^2}{2I}. \tag{4}$$

The ground state energy, $E_0 = 0$. The other energy levels are doubly degenerate, $E_m = E_{-m}$.

A periodic potential is often encountered for a hindered internal rotation. The n-fold potential energy can be written as,

$$V(\tau) = \frac{V}{2}(1 - \cos n\tau). \tag{5}$$

The time-independent Schrödinger equation is,

$$-\frac{\hbar^2}{2I}\frac{d^2\Phi}{d\tau^2}+\frac{V}{2}(1-\cos n\tau)\Phi = E\Phi. \tag{6}$$

Eq. 6 is a Mathieu-type equation. With the following definitions,

$$\begin{aligned} 2\phi &= n\tau, \\ a &= \frac{8I}{n^2\hbar^2}(E-\frac{V}{2}), \\ q &= -\frac{2V}{n^2}\frac{I}{\hbar^2}, \end{aligned} \tag{7}$$

Eq. 6 is transformed to the standard form of the Mathieu equation, Eq. 8,

$$\frac{d^2\Phi}{d\phi^2}+(a-2q\cos 2\phi)\Phi = 0. \tag{8}$$

For its significance in physics and other subjects, the Mathieu equation has been well studied. The eigenfunctions can be expanded on Fourier series. The even and odd eigenfunctions are usually written as, $\mathrm{ce}_m(\phi, q)$ and $\mathrm{se}_m(\phi, q)$, respectively, which have a 2π period in ϕ. They have been implemented as built-in functions in *Mathematica* along with the eigenvalues, a_m and b_m with m an integer or rotational number. The periodic boundary conditions are,

$$\Phi(2\phi/n) = \Phi(2\phi/n+2k\pi),\ \Phi(\phi) = \Phi(\phi+n\pi). \tag{9}$$

From Eq. 9, it is required that the eigenfunction $\Phi(\phi)$ has a period of $n\pi$ in ϕ.

It is common to find a three-fold symmetry potential energy curve for methyl and *t*-butyl groups. The eigenfunctions and eigenvalues are presented here. The symmetry is $n = 3$ in Eq. 6. If we use m to label the internal rotational quantum number, the normalized wave functions of Eq. 6 are given by Eqs. 10, 11 and 12. The ground state wave function, $m = 0$, is given by,

$$\Phi_0(\tau) = \frac{1}{\sqrt{2\pi}}\mathrm{ce}_0(\frac{3\tau}{2}, q), \tag{10}$$

with the characteristic value, $a_0(q)$.

When $m = 3, 6, 9, \ldots$, there are two states given by the odd and even Mathieu functions, which are A-type states in the C_3 group.

$$\Phi_{m,a}(\tau) = \frac{1}{\sqrt{2\pi}}\mathrm{ce}_{\frac{2m}{3}}(\frac{3\tau}{2}, q),$$

with the characteristic value, $a_{2m/3}(q)$,

$$\Phi_{m,b}(\tau)=\frac{1}{\sqrt{2\pi}}\mathrm{se}_{\frac{2m}{3}}(\frac{3\tau}{2},q), \tag{11}$$

with the characteristic value, $b_{2m/3}(q)$.

When m are other values, *i.e.* 1, 2, 4,5, 7, ..., there are two degenerated -type states given by the Mathieu functions,

$$\Phi_{m,+}(\tau)=\frac{1}{\sqrt{\pi}}(\mathrm{ce}_{\frac{2m}{3}}(\frac{3\tau}{2},\mathrm{q})+i\,\mathrm{se}_{\frac{2m}{3}}(\frac{3\tau}{2},\mathrm{q})), \tag{12}$$

and,

$$\Phi_{m,-}(\tau)=\frac{1}{\sqrt{\pi}}(\mathrm{ce}_{\frac{2m}{3}}(\frac{3\tau}{2},q)-i\,\mathrm{se}_{\frac{2m}{3}}(\frac{3\tau}{2},q)),$$

with the degenerated eigenvalues, $a_{2m/3}(q) = b_{2m/3}(q)$.

The energies of each state can be obtained from the corresponding characteristic values, $a_{2m/3}(q)$ and $b_{2m/3}(q)$, by applying the transformation rules in Eq. 7. The ascending ladder of energies is in the order of *A*, *E*, *E*, *E*, *E*, *A*, and this pattern is repeated. In the literature, other schemes are commonly used to label the rotational or torsional quantum number. Usually, the first three states are indexed with 0, next three states with 1 and so on. For the low-lying states, the energy difference between the *A* state and the degenerated *E* states is small compared with the gap to the next higher three *E* and *A* states. This *A*-*E* splitting is related to the quantum tunneling of the rotor at very low temperature [1,17,18].

The angular momentum operator for the internal rotation is $i\hbar\frac{d}{d\tau}$. From the wave functions, we may evaluate the expectation value of the angular momentum for each state,

$$<\Omega>=i\hbar\int_0^{2\pi}\Phi_m^*(\tau)\frac{d\Phi_m(\tau)}{d\tau}d\tau. \tag{13}$$

If we use the $\mathrm{ce}'_m(\phi, q)$ and $\mathrm{se}'_m(\phi, q)$ to represent the first derivative of the $\mathrm{ce}_m(\phi, q)$ and $\mathrm{se}_m(\phi, q)$ with respect to ϕ, which are also built-in functions in *Mathematica*, the expectation values can also be calculated. For the *A*-type states, $<\Omega> = 0$ since the $\mathrm{ce}_m(\phi, q)$ and the $\mathrm{ce}'_m(\phi, q)$ are orthogonal. The two degenerated *E*-type states have the same expectation values of the angular momentum but with opposite signs. For the states well-above the barrier height, the rotor behaves like a free-rotor with a shifted zero point energy. The angular momentum expectation values are proportional to the quantum number of the states.

$$<\Omega>=\pm\hbar\frac{2}{3\pi}\int_0^{2\pi}\mathrm{ce}_{\frac{2m}{3}}(\frac{3\tau}{2},q)\mathrm{se}'_{\frac{2m}{3}}(\frac{3\tau}{2},q)\mathrm{d}\tau,\ m\,\square\,1,2,3,4,5\cdots \tag{14}$$

A more general potential energy curve may be expanded with several Fourier terms. The corresponding Schrödinger equation can be solved with the perturbation method, the finite-difference-boundary-value method or other numerical methods [22-23]. In the later section **A.5**, the finite-difference-boundary-value method is also implemented besides the solutions expressed with the Mathieu functions. In addition, the equation is also solved to expand the wave functions in the free rotor basis set. Each method has its advantages and shortcomings. The four-dimensional rotational dynamics of the *t*-butyl group can be reduced into a two-dimensional problem. The Schrödinger equation can be solved by expanding the wave functions in the tensor product of two one-dimensional free rotor basis sets (one is for the overall *t*-butyl rotation and the other for the constituent methyl rotation) [7-9].

A.3. Reorientation Rate of Hindered Rotor

There are different models to calculate the rotational or the reorientation rates for the rotor groups hopping from one of the potential minima to another. In the high temperature limit, the classical rotational model is more convenient. In the low temperature region, tunneling occurs and a quantum mechanics treatment is required [1]. For the three-fold potential energy curve, there are three potential wells. At the low temperature limit, $V >> RT$, only the ground state is significantly populated. The bottom of the three potential wells has the largest probability for finding the orientation of the group. The ground state is *A*-symmetry and the expectation value of the angular momentum is 0 as discussed in the last section. The only process for reorientation is tunneling. The rate of tunneling among three wells, called the tunneling frequency, can be obtained by time-dependent perturbation theory. For a three-fold rotor, the tunneling frequency, ν_t, is given by

$$\nu_t = \Delta E/h \tag{15}$$

where ΔE is the so-called *A-E* splitting, the energy difference between the ground state and the first excited rotational state, which is an *E*-symmetry state.

The quantum tunneling dynamics of the methyl group rotation studied by nuclear magnetic resonance spectroscopy (NMR) has been reviewed in the reference [1]. In the high-temperature region, the classical hopping model works well, in which the reorientation is treated as an activated process. With an activation energy of E_a, the Arrhenius equation for the hopping process is written as,

$$\nu_c = \nu_0 e^{-E_a/RT}, \tag{16}$$

where T is temperature and ν_0 is the extrapolated value of ν_c to $1/T = 0$ from the experimental measured dependence of ν_c on T. The value of ν_0 for the methyl group is found to be of the order of 10^{12} s^{-1}. The activation barrier E_a is the potential well height V in classical mechanics. In quantum mechanics, the energy levels are discrete.

Assuming a Boltzmann distribution for the energy level population, the rotational rate dependence on temperature is written as,

$$\nu_q = \sum_m \nu_m e^{-E_m/RT} / \sum_m e^{-E_m/RT}, \tag{17}$$

where ν_m is the temperature independent rotational rate for the energy level, E_m. There are several models to treat the reorientation rate in solids in a full temperature range. The semi-classical model proposed by Clough et al.[18] is reviewed here. In Clough's model, the dynamics of the torsional and reorientation motion of the internal rotational group is separated from the lattice. The coupling between the lattice and the rotor maintains thermal equilibrium. Regarding the hopping process as a wave packet propagation, the rotation rate is calculated by equating the expectation value of the angular momentum $<\Omega> = I\omega = I2\pi\nu/3$. Therefore, the rotational rate ν_m for the m-state is written as,

$$\nu_m = \frac{3}{2\pi I}\left|<\Omega>\right|, \tag{18}$$

where the modulus removes the discrimination between the clockwise and anticlockwise rotations. At the temperature T, the energy levels of the rotor are populated by the Boltzmann distribution. The thermal averaged rotation rate is written as,

$$<\nu_m>_T = \frac{3}{2\pi I} <\left|<\Omega>\right|>_T, \tag{19}$$

where $<|<\Omega>|>_T$ is the thermal average value of the angular momentum. The hopping rate ν_h in the classical model is related to the parameter $<\nu_m>_T$, by,

$$\nu_h = <\nu_m>_T/2. \tag{20}$$

The half-width ν of the spectral density function associated with reorientation is connected with ν_h by the following equation,

$$\nu = 3\nu_h. \tag{21}$$

By combining Eqs. 18, 19 and 20, we obtain

$$\nu = \frac{3}{2}\frac{3}{2\pi I} <|< \Omega >|> \tau. \tag{22}$$

This model is found to be consistent with experimental observations in a wide temperature range. For the low-temperature regime, Eq. 18 needs to be modified to consider the quantum tunneling motion. Other models are discussed elsewhere [24].

A.4. Applications in Internal Rotations of Methyl, Trifluoromethyl and t-Butyl Groups

In this section, we apply the above-discussed theoretical methods to solving the Schrödinger equation of the three three-fold rotational groups. We will assume a three-fold potential energy curve in the form of Eq. 5 and use only one parameter, the barrier height, *V*, to characterize the potential energy. The moments of inertia of three groups are 3.09, 86.1 and 106.1 amu Å^2 for the methyl, trifluoromethyl and *t*-butyl groups, respectively. The variation of the inertia moments during the rotational processes is negligible.

In Figure A.1, the potential energy (black curve), energy levels below the rotational barrier (horizontal lines) and *A*-type state wave functions (blue curves) are plotted for the methyl group with an intermediate barrier height of 0.66 kcal/mol. The *A*-*E* tunnel splitting in the ground state is very small, about 0.3 cal/mol. And the splitting increases as the energy levels increase. For the energy levels above the barrier height, which are double-degenerated, the rotor behaves like a free rotor but the zero energy scale is shifted up by *V*/2. For a high barrier height, *e.g.* 3.0 kcal/mol, most of the thermally attainable energy levels at room temperature are below the barrier height. For the states near the bottom of the potential wells, the potential energy can be approximated by the first term of the Malcaurin series of $V(\tau)$,

$$V(\tau) = \frac{9V\tau^2}{4}. \tag{23}$$

The Schrödinger equation reduces to that of a harmonic oscillator. And the energy levels of the *A* and *E* states become degenerated.

Compared with the methyl group, the trifluoromethyl and *t*-butyl groups are much heavier and the moments of inertia are much larger. The density of rotational states per energy unit is more concentrated for the trifluoromethyl and *t*-butyl groups than for the methyl group. The rotor behaves more like a classical one and the dynamics could be well described with the Arrhenius equation in a wide temperature range. In Figure A.2, the energy levels are plotted for the t-butyl group at three barrier heights, assuming a three-fold cosine potential.

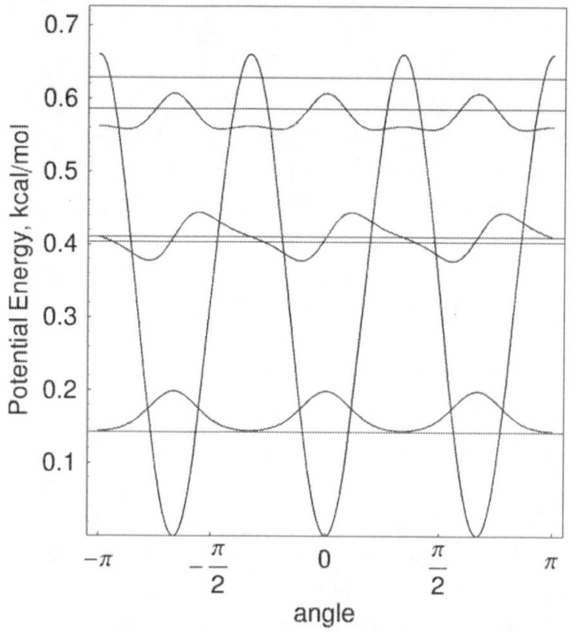

Figure A.1. Methyl Group: Potential energy curve, energy levels (horizontal lines) and *A*-type wave functions (blue curves).

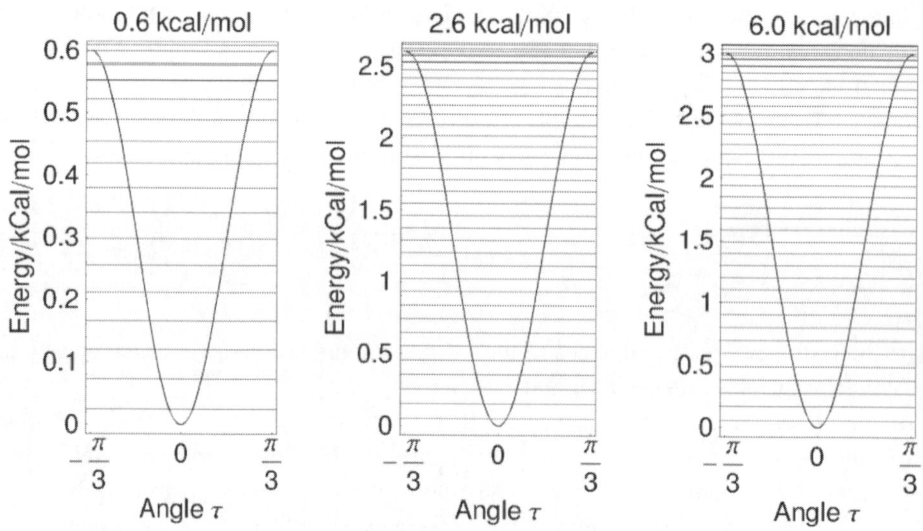

Figure A.2. *t*-Butyl Group: Three-fold cosine potential energy curves with rotational energy levels (horizontal lines) at three barrier heights, 0.6, 2.6 and 3.0 kcal/mol.

To investigate the preferential orientation of the groups, the thermally averaged distribution of the probability density (Φ*Φ) (with the Boltzmann weight) is calculated at 298 K for a few typical three-fold barrier heights. They are plotted in Figures A.3, A.4 and A.5. For the low barrier heights, *e.g.* 0.6 kcal/mol or lower, the distribution is widely distributed over all the angles. If the barrier is much higher than *RT*, the distribution is much narrower with the preferential orientation at the potential well. The half-width-at-half-maximum-height (HWHM) value for each distribution peak of the *t*-butyl group with 6.0 kcal/mol barrier height is about 8°. Therefore, at 298 K, there is a significant probability of finding the orientation of the *t*-butyl group within ±8° from the equilibrium orientation at the bottom of the potential well. The moment of inertia of the groups affects the distribution slightly. The HWHM value for the methyl group is a few degrees larger than that of the *t*-butyl group at the same barrier height. Trueblood [25] evaluated the possibility of deriving the barrier from the angular distribution of rotor groups in the crystals.

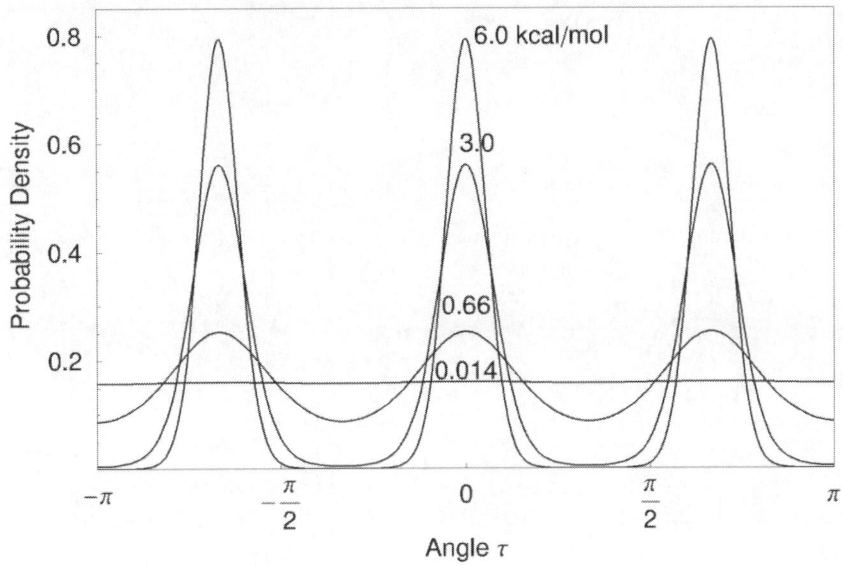

Figure A.3. Methyl Group: Probability density dependence on the orientation of the methyl group at 298 K with the three-fold barrier heights indicated on the graph.

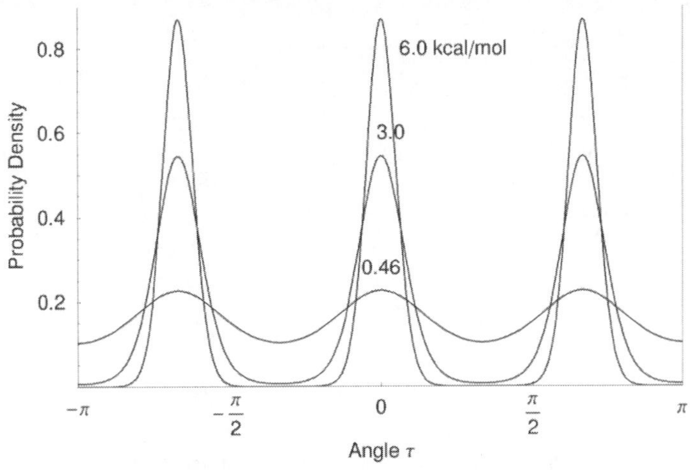

Figure A.4. Trifluoromethyl Group: Probability density dependence on the orientation of the trifluoromethyl group at 298 K with the three-fold barrier heights indicated on the graph.

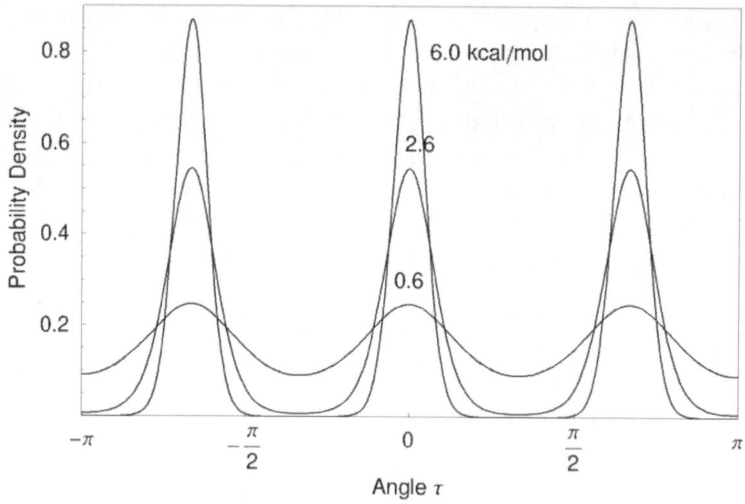

Figure A.5. *t*-Butyl Group: Probability density dependence on the orientation of the *t*-butyl group at 298 K with the three-fold barrier heights indicated on the graph.

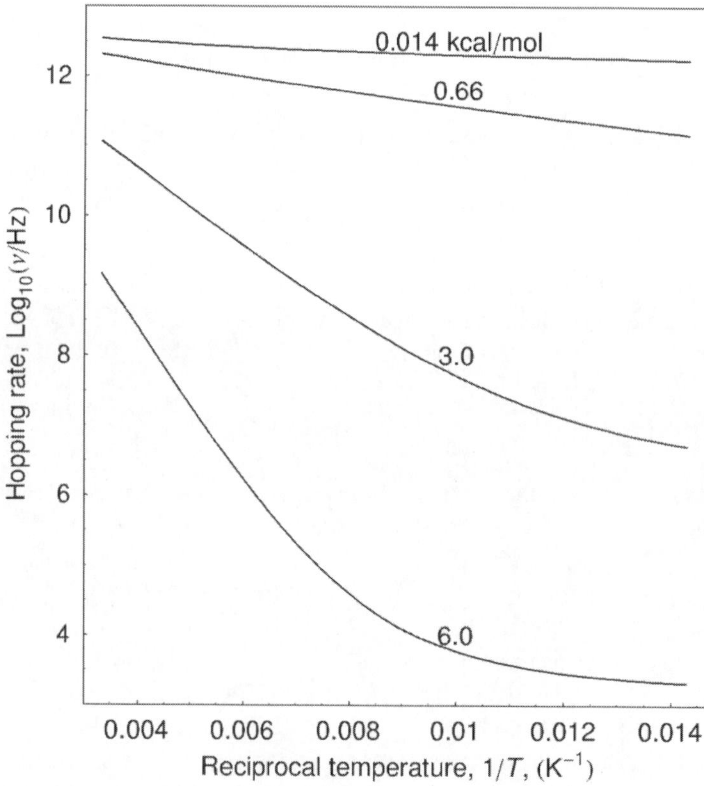

Figure A.6. Methyl Group: Hopping rate dependence on temperature in the region from 70 to 300 K with the three-fold barrier heights as indicated.

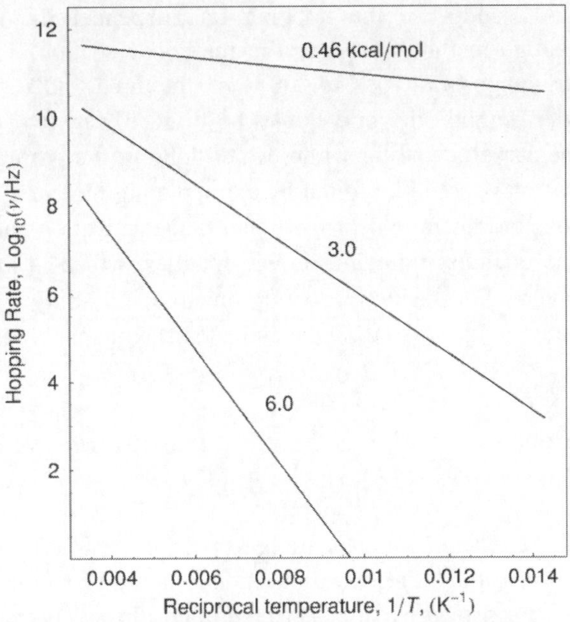

Figure A.7. Trifluoromethyl Group: Hopping rate dependence on temperature in the region from 70 to 300 K with the three-fold barrier heights as indicated.

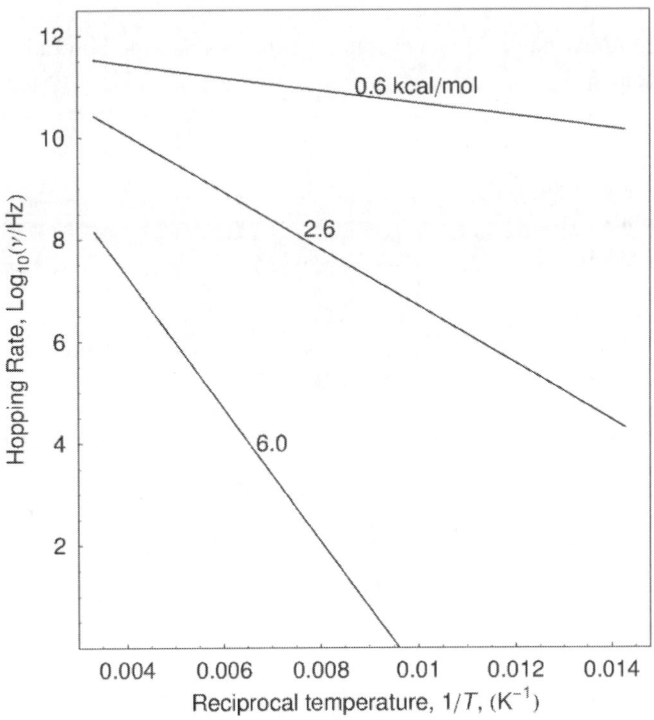

Figure A.8. *t*-Butyl Group: Hopping rate dependence on temperature in the region from 70 to 300 K with the three-fold barrier heights as indicated.

In Figures A.6, A.7 and A.8, the hopping rate dependences on temperature were calculated with the Clough method introduced in the above section. The calculations were done in the temperature range from 70 K to 300 K for the three-fold barriers. For the methyl group with high barrier heights, the curve may be divided into two regions. In the high temperature region, the curve is a straight line, which follows the Arrhenius's law of Eq. 16. The pre-exponential factor ν_0 could be found by extrapolating the curve to $1/T = 0$, which is about 10^{12} Hz. The slope of the curve is proportional to the activation energy, E_a. For the low temperature region, the quantum tunneling effect emerges, which is to lower the apparent activation energy E_a and ν_0. For the heavier trifluoromethyl and *t*-butyl groups, the curves are straight in the temperature range studied. The classical Arrhenius model fits well in the range.

A.5. Solution of Internal Rotation Schrödinger Equations: Mathematica Codes

In this section, the *Mathematica* codes are reported for the solutions of the Schrödinger equations of the one-dimensional and two-dimensional internal rotations. The energy levels and the wave functions are solved by three different algorithms. Thermal population is also treated.

A.5.1. Units

It is convenient to use natural units as listed below to solve the Schrödinger equation.

Angular Moment, $\hbar$;
Length, [Å];
Mass, [amu];
Moment of Inertia, [amu Å^2];
Energy, [$\hbar^2$/(amu Å^2)];
Frequency, [$\hbar$ /(amu Å^2)].

```
(*Energy in Natural Unit*)
nuE = (1.05457*10^-34 Joule \ Second)^2/(1.66054*10^-27 Kilogram (1.0*10^(-10) Meter)^2) 6.02214*10^23 \ Mole^-1 (403.324 Joule^2 Second^2)/(Kilogram Meter^2 Mole)
(*Conversion Constants From Natural Units to Regular Ones*)
nuEtoKJperMol = 0.403324;
nuEtoKCalperMol = nuEtoKJperMol/4.184;
nuEtoKelvin = nuEtoKJperMol*1000/8.314;
nuEtoWaveNumber = nuEtoKCalperMol*349.755;
nuFtoHz = 1.05457*10^-34/(1.66054*10^-27 (1.0*10^(-10))^2);
```

A.5.2. Free Rotor

The energy levels of a free rotor are given by $m^2/(2I)$ and the wave functions are in the form of $\frac{1}{\sqrt{2\pi}}e^{im\tau}$.

```
(*Wave functions and Energy Levels of Free Rotors*)
wavefunction[m_, τ_] := 1/Sqrt[2 π] E^(I m τ);
ParametricPlot3D[{τ, wavefunction[0, τ] // Re, wavefunction[0, τ] //
Im}, {τ, 0, 2 π}];
ParametricPlot3D[{τ, wavefunction[1, τ] // Re, wavefunction[1, τ] //
Im}, {τ, 0, 2π}];
ParametricPlot3D[{τ, wavefunction[2, τ] // Re, wavefunction[2, τ] //
Im}, {τ, 0, 2 π}];
ListPlot[Table[{i, i^2/(2 10) nuEtoKJperMol}, {i, 0, 100}]];
```

A.5.2. Periodic Potential Rotor

A n-fold potential of one-dimensional rotor is in the form of $V(\tau)=V/2\ (1\text{-}\cos(n\tau))$.

(i) Conversion To Standard Form Mathieu Equation:
With the following substitutions, the one-dimensional rotational Schrödinger equation hindered by the n-fold cosine potential can be converted into a Mathieu equation.
Substitutions:

$$2\phi = n\tau \ ; \ a = \frac{8I}{n^2}(E - \frac{V}{2}) ; \ 2q = -\frac{8I}{n^2}\frac{V}{2}.$$

Standard Form:

$$\frac{d^2\Phi}{d\phi^2} + (a - 2q\cos 2\phi)\Phi = 0.$$

(ii) Moment of Inertia
Methyl Group: 3.094 amu Å^2;
Trifluoromethyl Group: 86.14 amu Å^2;
t-Butyl Group: 106.1 amu Å^2.

(iii) Analytic Mathieu Solution
The following function **PeriodicPotenialRotor[]** returns the energy values and the wave functions for the rotational states.

```
(* INPUT:
```

```
symmetry - the n-fold symmetry number;
inertia - the moment of inertia in natural unit;
barrier - the barrier height in natural unit;
levels - the number of low-lying states requested.
OUTPUT:
The list of {energy, wavefunction, momentumA, momentumE} *)
PeriodicPotenialRotor[symmetry_, inertia_, barrier_, levels_] :=
Module[{i, q, f, energy, wavefunction, momentum},
f = 8 inertia/symmetry^2;
q = f *barrier/4;
energy = {};
wavefunction = {};
momentumA = {};
momentumE = {};
AppendTo[energy, MathieuCharacteristicA[0, q]/f + barrier/2];
AppendTo[momentumA, {0, MathieuCharacteristicA[0, q]/f +barrier/2}];
AppendTo[wavefunction, {MathieuC, 0, q, symmetry}];
Do[{
AppendTo[energy, MathieuCharacteristicA[(2/symmetry) i, q]/f +
barrier/2];
AppendTo[wavefunction, {MathieuC, 2 i/symmetry, q, symmetry}];
AppendTo[energy, MathieuCharacteristicB[(2/symmetry) i, q]/f +
barrier/2];
AppendTo[wavefunction, {MathieuS, 2 i/symmetry, q, symmetry}];
If[Mod[i, symmetry] == 0,
{momentum = 0;
AppendTo[momentumA, {momentum, MathieuCharacteristicA[(2/symmetry) i,
q]/f + barrier/2}];
momentum = 0;
AppendTo[momentumA, {i, MathieuCharacteristicB[(2/symmetry) i, q]/f +
barrier/2}]},
{momentum = 0;
AppendTo[momentumE, {momentum, MathieuCharacteristicA[(2/symmetry) i,
q]/f + barrier/2}]}]},
{i, levels}];
Return[{energy , wavefunction, momentumA, momentumE}];
];
result1 = PeriodicPotenialRotor[3, 86.14 , 12.4 /nuEtoKJperMol, 100];
```

(iv) Finite-Difference-Boundary-Value Solution

The following function **FiniteBoundarySol[]** solves the Schrödinger equation using the numeric finite difference boundary value method. The method can handle the potential in any arbitrary form. Here the potential is given by the function, **barrier3F[]**, which can be replaced by what is appropriate.

```
<<LinearAlgebra`MatrixManipulation`;
(*Potential Function Definition*)
barrier3F[bHeight_, \[Phi]_]:=bHeight (1- Cos[3\[Phi]])/2;
(* INPUT:
inertia - the moment of inertia in natural unit;
barrier - the barrier height;
size - to determine the accuracy of the solutions. Higher is better.
OUTPUT:
The list of {eigen_values, eigen_functions}. *)
```

```
FiniteBoundarySol[inertia_,barrier_,size_]:=Module[{n,s,fun,matrixF,r
esult},
n=size;
s= 2 Pi/n;
fun[a_,b_]:=If[a==b,barrier3F[barrier,-Pi  +  a*  s]+1/(inertia*s^2),-
1/(2inertia*s^2)];
matrixF=TridiagonalMatrix[fun,n];
matrixF=ReplacePart[matrixF,-1/(2inertia*s^2),{{1,n},{n,1}}];
result=Eigensystem[matrixF];
Return[{result[[1]],result[[2]]/Sqrt[s]}]
];
result=FiniteBoundarySol[1.57,12.4/nuEtoKJperMol,144];
result[[1]]//ListPlot;.
```

(v) Free Rotor Expansion Solution

The following function **FreeRotorExpansion[]** solves the three-fold cosine potential rotation equation using the free rotor wave functions as the expansion basis set.

```
<<LinearAlgebra`MatrixManipulation`;
 (*Three-fold Rotor: symmetry=3, component=-1,0,1 *)
FreeRotorExpansion[symmetry_,component_,inertia_,barrier_,size_]:=Mod
ule[{matrixE,result,fun},
fun[i_,j_]:=If[i==j,(component-  symmetry  size  +symmetry(i-1))^2/(2
inertia)+barrier/2,-barrier/4];
matrixE=TridiagonalMatrix[fun,2 size+1];
result=Eigensystem[matrixE];
Return[{Reverse[result[[1]]],Reverse[result[[2]]]}];
];
barrier=3.0;
result1=FreeRotorExpansion[3,1,106.1,barrier /nuEtoKCalperMol,200];
result1[[1]]nuEtoKCalperMol//ListPlot;
wcurve1=Plot[barrier  (1-Cos[3  x])/2,{x,  -Pi/3,Pi/3}  ,AspectRatio->
2,Frame->     True,FrameLabel->     {"Angle     τ","Energy/kCal/mol","6.0
kcal/mol",""},GridLines-
>{None,Union[result1[[1]],result2[[1]]]nuEtoKCalperMol},          Axes->
False,FrameTicks->{{-π,-(π/3),0,π/3,π},Automatic,  None,None},TextStyle->
{FontFamily->"Arial",FontSize->12}];
```

(vi) Thermodynamics

The following codes deal with the thermal population on the rotational states using the Boltzmann distribution. Angular distribution can be obtained. The Clough model is also used to calculate the dependence of the rotational rate on temperature. Other statistical properties can also be easily derived as well.

```
(*Construct Wavefunction From Mathieu Function Solution*)
WaveFuncFromMathieu[temp_] := Switch[Part[temp, 1],
MathieuC,
Return[(1/Sqrt[2  Pi])  MathieuC[MathieuCharacteristicA[Part[temp,  2],
Part[temp, 3]], Part[temp, 3], Part[temp, 4] (Pi - x)/2]],
MathieuS,
Return[(1/Sqrt[2  Pi])  MathieuS[MathieuCharacteristicB[Part[temp,  2],
Part[temp, 3]],Part[temp, 3], Part[temp, 4] (Pi - x)/2]]];
(*Construct Wavefunction From The Free Rotor Basis Solution*)
```

```
WaveFuncFromFreeRotor[temp_, symmetry_, component_] := Sum[temp[[i]] Cos[ (component - symmetry (Length[temp] - 1)/2 + symmetry (i - 1)) x], {i, Length[temp]}]/Sqrt[2 Pi];
(*Squared Wavefunction*)
WaveSquared[temp_, symmetry_] := Sum[If[i == j, temp[[i]]^2/(2 Pi), (temp[[i]] temp[[j]]/Pi) Cos[ symmetry (i - j) x]], {i, Length[temp]}, {j, i}];
MomentumExp[temp_, symmetry_, component_] := Sum[(component - symmetry (Length[temp] - 1)/2 + symmetry (i - 1)) temp[[i]]^2, {i, Length[temp]}];
(*Thermal Population on Rotational States a at Temperature t *)
ThermPop[a_, t_] := Module[{i, q, result},
q = Sum[Exp[-Part[a, 1, i] nuEtoKJperMol 1000/(8.314 t)], {i, Length[a[[1]]]}];
result = Table[Exp[-Part[a, 1, i] nuEtoKJperMol 1000/(8.314 t)]/q, {i, Length[a[[1]]]}];
Return[result];
];
(*Thermal Population on Free Rotor Basis Solution States a at Temperature t *)
ThermPopFreeRotorBasis[a_, t_] := Module[{i, j, q, result},
q = Sum[Exp[-Part[a, i, 1, j] nuEtoKJperMol 1000/(8.314 t)], {i, Length[a]}, {j, Length[Part[a, i, 1]]}];
result = Table[Exp[-Part[a, i, 1, j] nuEtoKJperMol 1000/(8.314 t)]/q, {i,Length[a]}, {j, Length[Part[a, i, 1]]}];
Return[result];
];
(*Angular Distribution at Temperature t *)
AngularDistribution[a_, t_] := Module[{i, p, dist},
p = ThermPop[a, t];
dist = Table[0, {x, -Pi, Pi, Pi/144}];
Do[dist = dist + p[[i]] Table[ContructWaveFunction[Part[a, 2, i]]^2, {x, -Pi, Pi, Pi/144}], {i, Length[a[[2]]]}];
dist = Transpose[{Table[x, {x, -Pi, Pi, Pi/144}], dist}];
Return[dist];
];
(*Angular Distribution at Temperature t for the States Obtained in \
Free Rotor Basis Method*)
AngularDistFreeRotorBasis[a_, t_, symmetry_] := Module[{i, p, dist, result},
p = ThermPopFreeRotorBasis[a, t];
result = Sum[Part[p, i, j] WaveSquared[Part[a, i, 2, j], symmetry], {i, Length[a]}, {j, Length[Part[a, i, 2]]}];
Return[result];
];
(*Average Angular Momentum at Temperature t*)
MomentumAverage[a_, t_, symmetry_] := Module[{i, p, dist, result},
p = ThermPopFreeRotorBasis[a, t];
result = 2 Sum[Abs[Part[p, i, j] MomentumExp[Part[a, i, 2, j], symmetry, 1]], {i,1}, {j, Length[Part[a, i, 2]]}];
Return[result];
];

(*Example*)
```

```
numb = 100;
inertia = 106.1;
barrier = 6.0;
result1 = FreeRotorExpansion[3, 1, inertia, barrier /nuEtoKCalperMol, 100];
result2 = FreeRotorExpansion[3, -1, inertia, barrier /nuEtoKCalperMol, 100];
result3 = FreeRotorExpansion[3, 0, inertia, barrier /nuEtoKCalperMol, 100];
ListPlot[Take[result3[[1]], numb] nuEtoKJperMol];
AngularFunc = AngularDistFreeRotorBasis[
{{Take[result1[[1]], numb], Take[result1[[2]], numb]}, {Take[result2[[1]], numb], Take[result2[[2]], numb]}, {Take[result3[[1]], numb], Take[result3[[2]], numb]}}, 298, 3];
plot11 = Plot[AngularFunc, {x, -Pi, Pi}];
```

(vii)Two-Dimensional Rotation

The Schrödinger equation of the internal rotational degrees of freedom of a t-butyl group can be reduced to a two-dimensional rotation, in which the coupling of the two rotations is connected by the potential energy term. The two-dimensional rotor can be solved by expanding the wave functions in the tensor product of two one-dimensional free rotor basis sets. One is for the overall t-butyl rotation and the other is for the rotation of the constituent methyl group.

```
<< RealTime3D`;
(*Constrained Methyl Rotation Barrier Dependence on t-Butyl Orientation *)
barrier = {
{0, 6.61403},
{15, 6.14456},
{30, 4.946},
{45, 4.12227},
{60, 4.19098},
{75, 4.44228},
{90, 4.44134},
{105, 4.29268},
{120, 4.03},
{135, 4.12227},
{150, 4.946},
{165, 6.14456},
{180, 6.61403}};
barrier = Transpose[barrier];
barrier = {Join[barrier[[1]] Degree, -barrier[[1]] Degree], Join[barrier[[2]], barrier[[2]]]} // Transpose;
orgplot = ListPlot[barrier, PlotStyle -> {Hue[.66], PointSize[.02]}]
(*The Dependence Relationship Can Be Expanded Using Cosine Potentials*)
fitResult = Fit[barrier,
{1, 1 - Cos[x], 1 - Cos[2 x], 1 - Cos[3 x], 1 - Cos[4 x], 1 - Cos[5 x], 1 - Cos[6 x]},x]
fitplot = Plot[fitResult, {x, -Pi, Pi}];
Show[orgplot, fitplot]

mSize = 50;
```

```
    nSize = 30;
    nBase = 0;
    (*Options: -1, 0, +1*)
    (*Moment of Inertia*)
    tButylInertia = 106.1;
    MetylInertia = 3.094;
    (*Potenial Function Parameters*)
    tButylBarrier = 1.0/nuEtoKCalperMol;
    MethylBarrier = {6.66, -0.018, -1.007, .0459, -.704, -.0361, -.0768}
\ 2/nuEtoKCalperMol;
    (*Hamitonian Matrix Element*)
    HamitonianMatrixElement[m0_, n0_, m1_, n1_] :=    KroneckerDelta[m0,
m1] m0^2 /(2 tButylInertia) +
    KroneckerDelta[n0, n1] n0^2 /(2 MetylInertia) +
    KroneckerDelta[n0, n1] (KroneckerDelta[m0, m1] tButylBarrier/2 -
KroneckerDelta[Abs[m0 - m1], 3] tButylBarrier/4) +
    KroneckerDelta[m0, m1] (KroneckerDelta[n0, n1] Part[MethylBarrier,
1]/2 - KroneckerDelta[Abs[n0 - n1], 3] Part[MethylBarrier, 1]/4) +
Sum[KroneckerDelta[n0, n1] (KroneckerDelta[m0, m1] Part[MethylBarrier,
k]/2 - KroneckerDelta[Abs[m0 - m1], k - 1] Part[MethylBarrier, k]/ 4) +
KroneckerDelta[Abs[n0 - n1], 3] (-1/2) (KroneckerDelta[m0, m1]
Part[MethylBarrier, k]/2 - KroneckerDelta[Abs[m0 - m1], k - 1]
Part[MethylBarrier, k]/ 4), {k, 2, 7}];
    (*Construct Hamiltonian Matrix*)
    (*Diagonalize the Matrix *)
    hMatrix = {};
    Do[Do[{temp = {},
    Do[Do[AppendTo[temp,
     HamitonianMatrixElement[m0, n0, m1, n1]], {m0, -mSize, mSize,
1}], {n0, -3 nSize + nBase, 3 nSize + nBase, 3}],
    AppendTo[hMatrix, temp]}, {m1, -mSize, mSize, 1}], {n1, -3 nSize +
nBase, 3 nSize + nBase, 3}];
    Print["Start Phase II..."];
    result = Eigensystem[hMatrix];
    ListPlot[result[[1]], PlotJoined -> True];
    ListPlot[result[[2]][[-2]], PlotJoined -> True, PlotRange -> All];
```

A.6. References

[1] Horsewill, AJ. Quantum tunneling aspects of methyl group rotation studied by NMR. *Prog. Nucl. Magn. Reson. Spectroscopy*, 1999, 35, 359-389.

[2] Koehler, JS; Dennison, DM. Hindered rotation in methyl alcohol. *Phys. Rev.*, 1940, 57, 1006-1021.

[3] Lin, CC; Swalen, JD. Internal rotation and microwave spectroscopy. *Rev. Modern Phys.*, 1959, 31(4), 841.

[4] Nielsen, HH. The torsion oscillator-rotator in the quantum mechanics. *Phys. Rev.*, 1932, 40, 445-456.

[5] Pottage, JT; Flower, DR; Davis, SL. The torsional excitation of methanol by helium. *J. Phys. B*, 2004, 37, 165-177.

[6] Spangler, LH. Structural information from methyl internal rotation spectroscopy. *Ann. Rev. Phys. Chem.*, 1997, 48, 481-510.

[7] Harthcock, MA; Laane, J. Calculation of two-dimensional vibrational potential energy surfaces utilizing prediagonalized basis sets and van Vleck perturbation methods. *J. Phys. Chem.*, 1985, 89, 4231-4240.

[8] Vivier-Bunge, A; Uc, VH; Smeyers, YG. Two-dimensional treatment of the methyl and aldehydic torsions in gas phase propanal. *J. Chem. Phys.*, 1998, 109(6), 2279-2286.

[9] Benderskii, VA; Makarov, DE; Grinevich, PG. Quantum chemical dynamics in two dimensions. *Chem. Phys.*, 1993, 170, 275-293.

[10] Bernstein, ER; Im, H.-S; Yong, MA. Supersonic Jet Studies of Ethoxybenzenes: Geometry of Their MinimumEnergy Conformations. *J. Org. Chem.*, 1991, 56, 6059-6063.

[11] Breen, PJ; Warren, JA; Bernstein, ER. Torsional motion in aromatic molecules. Conformational analysis of methyl-, ethyl-, and n -propylbenzenes. *J. Am. Chem. Soc.*, 1987, 109, 3453-3455.

[12] Disselkamp, R; Bernstein, ER; ISJ; Secor, HV. Minimum energy conformation of ortho-xylene in its ground and first excited electronic states. *J. Chem. Phys.*, 1992, 97, 8130-8136.

[13] Im, H-S; Bernstein, ER; Secor, HV; ISJ. Supersonic jet studies of benzyl alcohols: Minimum energy conformations and torsional motion. *J. Am. Chem. Soc.*, 1991, 113, 4422-4431.

[14] Seeman, JI; Secor, HV; Breen, PJ; Grassian, VH; Bernstein, ERA study of nonrigid aromatic molecules. Observation and spectroscopic analysis of the stable conformations of various alkylbenzenes by supersonic molecular jet laser spectroscopy. *J. Am. Chem. Soc.*, 1989, 111, 3140-3150.

[15] Seeman, J; Paine, JBI; Secor, HV; Im, H-S; Bernstein, ER. Supersonic jet studies of alkyl-substituted pyrazines and pyridines. Minimum energy conformations and torsional motion. *J. Am. Chem. Soc.*, 1992, 114, 5269-5280.

[16] Spangler, LH; Pratt, DW. Internal rotation dynamics from electronic spectroscopy in supersonic jets and beams. In *Jet spectroscopy and molecular dynamics*, Hollas, J. M. and Phillips, D., Eds. Blackie Academic & Professional: New York, 1995, 366-366.

[17] Allen, PS. A model for a temperature -dependent frequency distribution of methyl group tunneling splittings. *J. Phys. C: Solid State Phys.*, 1974, 7, L22-L25.

[18] Clough, S; Heidemann, A; Horsewill, AJ; Lewis, JD; Paley, MNJ. The rate of thermally activated methyl group rotation in solids. *J. Phys. C: Solid State Phys.*, 1982, 18, 2495-2508.

[19] Abragam, A. *The Principles of Nuclear Magnetism*. Oxford University Press: Oxford, 1961.

[20] Lipari, G; Szabo, A. Nuclear Magnetic resonance relaxation in nucleic acid fragments: models for internal motion. *Biochemistry*, 1981, 20, 6250-6256.

[21] Slichter, CP. *Principles of Magnetic Resonaces*. 3rd ed.; Springer: Berlin, 1990.

[22] Mellor, WE; Lee, AR; Kalotas, TM. Alternative calculations for internal rotations: Assessment via Mathieu and multi-Fourier term potentials. *Chem. Phys.*, 1997, 219, 257-263.

[23] Witschel, W; Hartwigsen, C. Partition function and the level density of the hindered rotor. *Chem. Phys. Lett.* 1997, 273, 304-310.

[24] Ushiyama, H; Takatsuka, K. Timee-dependent probability of quantum tunnelling in terms of the quasisemiclassical method. *J. Chem. Phys.* 2004, 120(10), 4561-4572 and references therein.

[25] Trueblood, KN; Dunitz, JD. Internal molecular motions in crystals. The estimation of force constants, frequencies and barriers from diffraction data. A feasibility study. *Acta Crystallographica B*, 1983, B39(1), 120-133.

INDEX

D

E

F

G

H

I

J

K

L

M